Compendium of Chemical Warfare Agents

Compendium of Chemical Warfare Agents

Steven L. Hoenig

Senior Chemist/Chemical Terrorism Coordinator
Florida Department of Health
Bureau of Laboratories-Miami

 Springer

Library of Congress Control Number: 2006926880

ISBN-10: 0-387-34626-0
ISBN-13: 978-0-387-34626-7

Printed on acid-free paper

9 8 7 6 5 4 3 2 1

springer.com

To Lena

Preface

It should be noted that this book, although it contains some materials from a previously published work, it is by no means the same. Each chemical agent listing has been expanded to include significant additional information. In particular, the listings now include sections containing Informational, Physical and Chemical Properties, Reactivity, Toxicity, Safety, and Military Significant Information. Although listed as Military Significant Information, this section can be used by any first responder, or any person in the field. It is not strictly for military use.

The purpose of this particular book is to equip those professionals in the federal, state, and local level organizations that are currently involved in Homeland Security with the necessary information to be able to address a chemical terrorist situation.

Those professionals would be from Emergency Operations Centers, HazMat Teams, Fire Departments, Health Departments, Civil Support Teams (CST), Military Personnels, various Intelligence Organizations, Clinical Laboratories, and Environmental Laboratories.

This would also include first responders such as emergency medical technicians, emergency room nurses, doctors, technicians, firefighters, police officers, clinicians, laboratorians, health department personnels, epidemiologists, as well as researchers and related professionals.

In addition, both private and public hospitals would need this information when those exposed to the suspected chemical terrorist agents are brought in for treatment.

Each chapter begins with a brief introduction about the particular class of agent, and then a brief paragraph about each individual agent.

Chapters 1 through 7 cover blister, blood, choking, incapacitating, nerve, tear, and vomit agents. There are four additional blister agents in Chapter 1, three additional choking agents in Chapter 3, and seven additional nerve agents in Chapter 5. It should be noted that the subject of Novichok compounds have been addressed in detail in Chapter 5. Do bear in mind that these compounds have been subjected to some debate, but very little is known about them in detail.

In addition, Chapter 6 has three additional tear agents, mixtures that are not included in the other book on chemical warfare.

Chapter 8 covers components that are used to make binary chemical weapons. The chapter covers three main components that are used to make GB2 and VX2 (the binary equivalents of GB and VX). Only those materials not easily found elsewhere are listed. For example, one of the components for binary VX2 is elemental sulfur (designated NE), the information about sulfur is readily available and is not dealt with herein.

Appendix A is a glossary of the terminology and abbreviations used throughout the book.

Appendix B covers some relevant chemical and physical concepts used in the book.

Appendix C is a quick reference chart for signs and symptoms of exposure to chemical terrorism agents. The chart is only meant as a guide and is not to be all-inclusive.

Appendix D contains the FTIR spectra of the five listed nerve agents.

Appendix E is a quick cross-reference among the chemical agent, symbol, type of agent, and CAS number.

Appendix F is a list of precursor chemicals used in the synthesis of typical chemical warfare agents.

Appendix G is the periodic table of elements.

I hope that all those who use this book find the information useful and practical. Suggestions are always welcomed.

Steven L. Hoenig

Acknowledgments

I wish to express my thanks to Richard Kolodkin, who has been a true friend and more over all of these years, and through all of the hardships.

Once again I express my thanks to Richard Leff, whose friendship I deeply appreciate and without whose assistance I would not be here now.

Contents

Chapter 1

Blister Agents

A ll of the blister agents (vesicants) are persistent, and all may be employed in the form of colorless gases and liquids. Blister agents damage any tissue that they come in contact with. They affect the eyes and lungs and blister the skin. They damage the respiratory tract when inhaled and cause vomiting and diarrhea when absorbed. Vesicants poison food and water and make other supplies dangerous to handle. They may produce lethalities, but skin damage is their main casualty producing effect. The severity of a blister agent burn directly relates to the concentration of the agent and the duration of contact with the skin. In addition to casualty production, blister agents may also be used to restrict use of terrain, to slow movements, and to hamper use of materiel and installations.

During World War I mustard (H) was the only blister agent in major use. It had a recognizable, distinctive odor and a fairly long duration of effectiveness under normal weather conditions. Since then, odorless blister agents have been developed that vary in duration of effectiveness. Most blister agents are insidious in action; there is little or no pain at the time of exposure. Exceptions are lewisite and phosgene oxime (CX), which cause immediate pain on contact. CX produces a wheal (similar to a bee sting) rather than a water blister, which the other blister agents produce.

Blister agents can be described as mustards, arsenicals, or urticants (chemical agents that causes itching or stinging). The mustards (H, HD, HN–1, HN–2, HN–3, Q, and T) contain either sulfur or nitrogen. The arsenical (ethyldichloroarsine [ED], methyldichloroarsine [MD], and phenyldichloroarsine [PD]) are a group of related compounds in which arsenic is the central atom. Arsenical's hydrolyze rapidly and are less toxic than other agents of military interest. Mustards and arsenical are sometimes mixed to alter their properties for military effectiveness; they may also be employed with thickeners.

1.1. Distilled Sulfur Mustard – HD

H and HD are blister and alkylating agents, producing cytotoxic action on the hematopoietic (blood-forming) tissues. The rate of detoxification of H and

HD in the body is very slow, and repeated exposures produce a cumulative effect. Its toxic hazard is high for inhalation, ingestion, and skin and eye absorption, but the most common acute hazard is from liquid contact with eyes or skin. Agent HD is distilled H, it has been purified by washing and vacuum distillation to reduce sulfur impurities. Agent H is a mixture of 70% bis-(2-chloroethyl) sulfide and 30% sulfur impurities produced by unstable Levinstein process.

Informational

Designation	HD
Class	Blister agent
Type	B – persistent
Chemical name	Bis-(2-chloroethyl) sulfide
CAS number	[505-60-2]

Chemical and Physical Properties

Appearance	Mustard agent *liquid* is colorless when pure, but it is normally a yellow to brown oily substance.
Odor	Mustard agent *vapor* is colorless with a slight garlic- or mustard-like odor.
Chemical formula	$C_4H_8Cl_2S$
Molecular weight	159.08
Chemical structure	$Cl-CH_2CH_2-S-CH_2CH_2-Cl$
Melting point	14.45°C
Boiling point	227.8°C (216°C calculated; decomposes)
Flash point	105°C
Decomposition temperature	180°C
Vapor density	5.4 (air = 1)
Liquid density	1.27 g/cm³ at 20°C
Solid density	1.338 g/cm³ at 13°C 1.37 g/cm³ at 0°C
Vapor pressure	0.072 mmHg at 20°C 0.11 mmHg at 25°C
Volatility	75 mg/m³ at 0°C (solid) 610 mg/m³ at 20°C (liquid) 2,860 mg/m³ at 40°C (liquid)

Solubility	Very sparingly soluble in water (less than 1%); freely soluble in oils and fats, gasoline, kerosene, acetone, carbon tetra-chloride, alcohol, PS, and FM. Miscible with DP, L, ED, PD and the organophosphorus nerve agents.

Reactivity

Hydrolysis products	Hydrochloric acid and thiodiglycol
Rate of hydrolysis	Half-life is 8.5 min in distilled water at 25°C and 60 min in salt water at 25°C. Mustard on or under water undergoes hydrolysis only if dissolved. It is only slightly soluble in water; as a result mustard may persist in water for long periods. Alkalinity and higher temperatures increase the rate of hydrolysis.
Stability	Stable at ambient temperatures; decomposition temperatures is 180°C; can be active for at least 3 years in soil; stable for days to week, under normal atmospheric temperature; slowly hydrolyzed by water; destroyed by strong oxidizing agents.
Storage stability	Stable in steel or aluminum containers. Rapidly corrosive to brass at 65°C; will corrode steel at 0.001 in./month at 65°C.
Decomposition	Mustard will hydrolyze to form hydrochloric acid and thiodiglycol.
Polymerization	Will not occur.

Toxicity

LD_{50} (skin)	100 mg/kg
LCt_{50} (respiratory)	1,000–1,500 mg-min/m^3
LCt_{50} (percutaneous)	10,000 mg-min/m^3
ICt_{50} (respiratory)	1,500 mg-min/m^3
ICt_{50} (percutaneous)	2,000 mg-min/m^3 or less. Wet skin absorbs more mustard than does dry skin. For this reason, HD exerts a casualty effect at lower concentrations in hot, humid weather, because the body is then moist with perspiration. The dosage given for skin absorption applies to temperatures of approximately 21–27°C, as the body would not be perspiring excessively at these temperatures. Above 27°C perspiration causes increased skin absorption. The incapacitating dose requirement drops rapidly as perspiration increases; at 32°C, 1,000 mg/m^3 could be incapacitating.
Rate of detoxification	Very low. Even very small, repeated exposures of HD are cumulative in their effects or more than cumulative owing to sensitization. This has been shown in the postwar case histories of workers in mustard-filling plants. Exposure to vapors from spilled HD causes minor symptoms, such as "red eye." Repeated exposure to vapor causes 100% disability

by irritating the lungs and causing a chronic cough and pain in the chest.

Skin and eye toxicity
Eyes are very susceptible to low concentrations; incapacitating effects by skin absorption require higher concentrations.

Rate of action
Delayed – usually 4–6 h until first symptoms appear. Latent periods of up to 24 h have been observed, however, and in rare cases of up to 12 days.

Overexposure effects
HD is a vesicant (blister agent) and alkylating agent producing cytotoxic action on the hematopoietic (blood-forming) tissues, which are especially sensitive. The rate of detoxification of HD in the body is very slow, and repeated exposures produce a cumulative effect. The physiological action of HD may be classified as local and systemic. The local action results in conjunctivitis or inflammation of the eyes, erythema which may be followed by blistering or ulceration; inflammation of the nose, throat, trachea, bronchi, and lung tissue. Injuries produced by HD heal much more slowly and are more susceptible to infection than burns of similar intensity produced by physical means or by most other chemicals. Systemic effects of mustard may include malaise, vomiting, and fever, with onset time about the same as that of the skin erythema. With amounts approaching the lethal dose, injury to bone marrow, lymph nodes, and spleen may result. HD has been determined to be a human carcinogen by the International Agency for Research on Cancer.

Safety

Protective gloves
Wear butyl toxicological agent protective gloves (M3, M4, glove set).

Eye protection
Wear chemical goggles as a minimum; use goggles and face shield for splash hazard.

Other
Wear gloves and lab coat with M9 or M17 mask readily available for general lab work. In addition, wear daily clean smock, foot covers, and head cover when handling contaminated lab animals.

Emergency procedures
Inhalation: Remove victim from the source immediately; administer artificial respiration if breathing has stopped; administer oxygen if breathing is difficult; seek medical attention immediately.

Eye Contact: Speed in decontaminating the eyes is absolutely essential; remove person from the liquid source, flush the eyes immediately with water by tilting the head to the side, pulling the eyelids apart with the fingers, and pouring water slowly into the eyes; do not cover eyes with bandages; but if necessary, protect eyes by means of dark or opaque goggles; seek medical attention immediately.

Skin Contact: Don respiratory protective masks and gloves; remove victim from agent source immediately; flush skin and clothes with 5% solution of sodium hypochlorite or liquid household bleach within 1 min; cut and remove contaminated clothing; flush contaminated skin area again with 5% sodium hypochlorite solution; then wash contaminated skin area with soap and water; seek medical attention immediately.

Ingestion: Do not induce vomiting; give victim milk to drink; seek medical attention immediately.

Military Significant Information

Field protection	Protective mask and permeable protective clothing for ED vapor and small droplets; impermeable protective clothing for protection against large droplets, splashes, and smears.
Decontamination	STB, fire, or DS2. Decontaminate liquid agent on the skin with the M258A1, M258, or M291 skin decontaminating kit. Decontaminate individual equipment with the M280 individual equipment decontamination kit.
Persistency	Depends upon the amount of contamination by liquid, the munition used, the nature of the terrain and the soil, and the weather conditions. Heavily splashed liquid persists 1–2 days or more in concentrations that produce casualties of military significance under average weather conditions, and a week to months under very cold conditions. HD on soil remains vesicant for about 2 weeks. An incident in which mustard leaked and soaked into soil caused blisters after 3 years. HD is calculated to evaporate about 5× more slowly than GB. Persistency in running water is only a few days, while persistency in stagnant water can be several months. HD is about twice as persistent in seawater.
Use	Delayed-action casualty agent.

1.2. Ethyldichloroarsine – ED

The Germans introduced ED in 1918 in an effort to produce a volatile agent with a short duration effectiveness that would act more quickly than diphosgene or mustard and that would last longer in its effects than PD. Like other chemical agents containing arsenic, ED is irritating to the respiratory tract and will produce lung injury upon sufficient exposure. The vapor is irritating to the eyes, and the liquid may produce severe eye injury. The absorption of either vapor or liquid through the skin in sufficient amounts may lead to systemic poisoning or death. Prolonged contact with either liquid or vapor blisters the skin.

Informational

Designation	ED
Class	Blister agent

Type	A – nonpersistent
Chemical name	Ethyldichloroarsine
CAS number	[598-14-1]

Chemical and Physical Properties

Appearance	Colorless liquid
Odor	Fruity but biting; irritating
Chemical formula	$C_2H_5AsCl_2$
Molecular weight	174.88
Chemical structure	$CH_3CH_2-As\diagup^{Cl}_{\diagdown Cl}$
Melting point	$< -65°C$
Boiling point	$156°C$
Flash point	No data available
Decomposition temperature	Stable to boiling point
Vapor density	6.0 (air = 1)
Liquid density	1.66 g/cm^3 at $20°C$
Solid density	No data available
Vapor pressure	2.09 mmHg at $20°C$ 15.1 mmHg at $50°C$
Volatility	6,500 mg/m^3 at $0°C$ 20,000 mg/m^3 at $20°C$ 27,200 mg/m^3 at $25°C$
Solubility	Soluble in ethyl chloride, alcohol, ether, benzene, acetone, and cyclohexane. Hydrolyzes immediately in the presence of water.

Reactivity

Hydrolysis products	Hydrochloric acid and ethylarsineous oxide.
Rate of hydrolysis	Rapid
Stability	No data available
Storage stability	Stable in steel. Attacks brass at $50°C$; destructive to rubber and plastics.
Decomposition	No data available
Polymerization	No data available

Toxicity

LD$_{50}$ (skin)	No data available
LCt$_{50}$ (respiratory) *LCt$_{50}$* *(percutaneous)*	3,000–5,000 mg-min/m^3, depending upon the period of exposure. Because the body detoxifies ED at an appreciable rate, the product of concentration and time is not a constant; as time increases, concentration does not decrease proportionately. For example, exposure to 40 mg/m^3 for 75 min might have an effect similar to that produced by exposure to 30 mg/m^3 for 166 min.
ICt$_{50}$ (respiratory)	5–10 mg-min/m^3 (temporary)
ICt$_{50}$ (percutaneous)	No data available
Rate of detoxification	Sublethal amounts detoxify rapidly, similar to other arsenicals.
Skin and eye toxicity	Vapor is irritating but not harmful to eyes and skin except on prolonged exposure. Liquid ED has about one-twentieth the blistering action of liquid L.
Rate of action	Irritating effect on nose and throat is tolerable after 1 min at moderate concentrations; blistering effect is less delayed than with HD, which may be delayed 12 h or longer.
Overexposure effects	No data available

Safety

Protective gloves	Wear butyl toxicological agent protective gloves (M3, M4, glove set).
Eye protection	Wear chemical goggles as a minimum; use goggles and face shield for splash hazard.
Other	Wear gloves and lab coat with M9 or M17 mask readily available for general lab work. In addition, wear daily clean smock, foot covers, and head cover when handling contaminated lab animals.
Emergency procedures	**Inhalation:** Remove victim from the source immediately; administer artificial respiration if breathing has stopped; administer oxygen if breathing is difficult; seek medical attention immediately. **Eye Contact:** Speed in decontaminating the eyes is absolutely essential; remove person from the liquid source, flush the eyes immediately with water by tilting the head to the side, pulling the eyelids apart with the fingers, and pouring water slowly into the eyes; do not cover eyes with bandages; but if necessary, protect eyes by means of dark or opaque goggles; seek medical attention immediately. **Skin Contact:** Don respiratory protective masks and gloves; remove victim from agent source immediately; flush skin and clothes with 5% solution of sodium hypochlorite or liquid

household bleach within 1 min; cut and remove contaminated clothing; flush contaminated skin area again with 5% sodium hypochlorite solution; then wash contaminated skin area with soap and water; seek medical attention immediately.

Ingestion: Do not induce vomiting; give victim milk to drink; seek medical attention immediately.

Military Significant Information

Field protection	Protective mask and permeable protective clothing for ED vapor and small droplets; impermeable protective clothing for protection against large droplets, splashes, and smears.
Decontamination	Not usually necessary in the field. If necessary for enclosed areas, use HTH, STB, household bleach, caustic soda, or DS2. Decontaminate liquid agent on the skin with the M258A1, M258, or M291 skin decontaminating kit. Decontaminate individual equipment with the M280 individual equipment decontamination kit.
Persistency	Short
Use	Delayed-action casualty agent

1.3. Lewisite – L-1

L-1 is a vesicant (blister agent); also, it acts as a systemic poison, causing pulmonary edema, diarrhea, restlessness, weakness, subnormal temperature, and low blood pressure. In order of severity and appearance of symptoms, it is a blister agent, a toxic lung irritant, absorbed in tissues, and a systemic poison. When inhaled in high concentrations, it may be fatal in as short a time as 10 min. L-1 is not detoxified by the body. Common routes of entry into the body include ocular, percutaneous, and inhalation.

Informational

Designation	L-1
Class	Blister agent
Type	B – persistent
Chemical name	2-Chlorovinyldichloroarsine
CAS number	[541-25-3]

Chemical and Physical Properties

Appearance	Pure Lewisite-1 is a colorless, oily liquid. The plant sample (war gas) is an amber to dark brown liquid.

Odor	Pure Lewisite-1 has with very little odor. The plant sample (war gas) has a geranium-like odor.
Chemical formula	$C_2H_2AsCl_3$
Molecular weight	207.32

Chemical structure

$$ClCH = CH - As \Big\langle \begin{matrix} Cl \\ Cl \end{matrix}$$

Melting point	−18°C (mix) 1°C (trans) −45°C (cis)
Boiling point	190°C (mix) 197°C (trans) 170°C (cis)
Flash point	None
Decomposition temperature	>100°C
Vapor density	7.1 (air = 1)
Liquid density	1.89 g/cm^3 at 20°C
Solid density	No data available
Vapor pressure	0.087 mmHg at 0°C 0.22 mmHg at 20°C 0.35 mmHg at 25°C
Volatility	1,060 mg/m^3 at 0°C 4,480 mg/m^3 at 20°C 8,620 mg/m^3 at 30°C
Solubility	Soluble in organic solvents and oils; insoluble in water and dilute mineral acids. Because of its good miscibility with other chemical warfare agents, L-1 is suitable for the preparation of tactical mixtures.

Reactivity

Hydrolysis products	Hydrochloric acid and chlorovinyl arsenous oxide, a vesicant. The latter is a nonvolatile solid that is not readily washed away by rains. Strong alkalies destroy these blister-forming properties.
Rate of hydrolysis	Rapid for vapor and dissolved Lewisite-1. Low solubility in water limits the hydrolysis.
Stability	Reasonably stable; however, in presence of moisture, it hydrolyses rapidly, losing its vesicant property.
Storage stability	Stable in steel or glass container. Corrosive to steel at a rate of 1×10^{-5} to 5×10^{-5} in./month at 65°C.
Decomposition	It hydrolyses in acidic medium to form HCl and nonvolatile (solid) chlorovinylarsenious oxide, which is a less potent vesicant than ED. Hydrolysis in alkaline medium, as in decontamination with alcoholic caustic or carbonate solution, produces acetylene and trisodium arsenate (Na_3ASO_4).

Therefore, decontaminated solution would contain toxic arsenic.

Polymerization   No data available

Toxicity

LD$_{50}$ (skin)   30 mg/kg

LCt$_{50}$ (respiratory)   1,400 mg-min/m^3. The intense irritation to the respiratory tract usually causes exposed personnel to mask immediately to avoid the vapor.

LCt$_{50}$ (percutaneous)   100,000 mg-min/m^3. When the humidity is high, L-1 hydrolyzes so rapidly that it is difficult to maintain a vapor concentration sufficient to blister bare skin. The high vapor pressure and short duration of effectiveness of L-1 further increases this difficulty.

ICt$_{50}$ (respiratory)   No data available

ICt$_{50}$ (percutaneous)   >1,500 mg-min/m^3. Lewisite irritates the eyes and skin and gives warning of its presence.

Rate of detoxification   The body does not detoxify Lewisite-1

Skin and eye toxicity   Even limited concentrations of L-1 vapor (below 300 mg-min/m^3) cause extreme irritation of the eyes. Burning, pain, sensitivity to light, tearing, and swelling of the eyes result. An exposure of 1,500 mg-min/m^3 produces severe and probably permanent corneal damage to the eyes. Liquids cause severe damage to the eyes. L-1 has about the same blistering action on the skin as HD, even though the lethal dosage for L-1 is much higher.

Rate of action   Rapid. The body absorbs L-1 more rapidly through the skin than the nitrogen mustards.

Overexposure effects   L-1 is a vesicant (blister agent); also, it acts as a systemic poison, causing pulmonary edema, diarrhea, restlessness, weakness, subnormal temperature, and low blood pressure. In order of severity and appearance of symptoms, it is a blister agent, a toxic lung irritant, absorbed in tissues, and a systemic poison. When inhaled in high concentrations, it may be fatal in as short a time as 10 min. L-1 is not detoxified by the body. Common routes of entry into the body include ocular, percutaneous, and inhalation.

Safety

Protective gloves   Wear Norton Chemical Protection Glove Set, M3 butyl rubber

Eye protection   Wear protective eyeglasses as a minimum; use goggle sand face shield for splash hazards.

Other   Wear full protective clothing (Level A) consisting of M3 butyl rubber suit with hood, M2A1 boots, M3 gloves, impregnated underwear, M9 series mask and coveralls (if desired), or the 30 mil Demilitarization Protective Ensemble (DPE) if available, or NIOSH-approved equivalent; wear gloves and lab coat with M9, M17, or M40 mask readily available for general lab work. In addition, wear daily clean smock, foot covers, and head covers when handling contaminated lab animals.

Emergency procedures

Inhalation: Remove from the source immediately; give artificial respiration if breathing has stopped; administer oxygen if breathing is difficult; seek medical attention immediately.

Eye Contact: Speed in decontaminating the eyes is absolutely essential; remove person from the liquid source; flush the eyes immediately with water for 10–15 min by tilting the head to the side, pulling eyelids apart with fingers, and pouring water slowly into the eyes; do not cover eyes with bandages, but if necessary, protect eyes by means of dark or opaque goggles; seek medical attention immediately.

Skin Contact: Remove victim from source immediately and remove contaminated clothing; immediately decon affected areas by flushing with 10% sodium carbonate solution; wash off with soap and water after 3–4 min to protect against erythema; seek medical attention immediately.

Ingestion: Do not induce vomiting; give victim milk to drink; seek medical attention immediately.

Military Significant Information

Field protection

Protective mask and permeable protective clothing for L-1 vapor and small droplets; impermeable protective clothing for protection against large droplets, splashes, and smears.

Decontamination

HTH, STB, household bleach, DS2, or caustic soda. Decontaminate liquid agent on the skin with the M258A1, M258, or M291 skin decontaminating kit. Decontaminate individual equipment with the M280 individual equipment decontamination kit.

Persistency

Somewhat shorter than for HD; very short duration under humid conditions.

Use

Moderately delayed-action casualty agent.

1.4. Lewisite – L-2

L-2 is a vesicant (blister agent); also, it acts as a systemic poison, causing pulmonary edema, diarrhea, restlessness, weakness, subnormal temperature, and low blood pressure. In order of severity and appearance of symptoms, it is a blister agent, a toxic lung irritant, absorbed in tissues, and a systemic poison. When inhaled in high concentrations, it may be fatal in as short a time as 10 min. L-2 is not detoxified by the body. Common routes of entry into the body include ocular, percutaneous, and inhalation.

Informational

Designation	L-2
Class	Blister agent
Type	B – persistent
Chemical name	Bis-(2-chlorovinyl)chloroarsine
CAS number	[40334-69-8]

Chemical and Physical Properties

Appearance	Colorless to brownish
Odor	Fruity or flowery odor
Chemical formula	$C_4H_4AsCl_3$
Molecular weight	233.4
Chemical structure	$$ClCH=CH-\overset{\overset{\textstyle Cl}{\textstyle |}}{As}-CH=CHCl$$
Melting point	No data available
Boiling point	230°C
Flash point	No data available
Decomposition temperature	No data available
Vapor density	No data available
Liquid density	1.70 g/cm³ at 20°C
Solid density	No data available
Vapor pressure	No data available
Volatility	No data available
Solubility	No data available

Reactivity

Hydrolysis products	Hydrochloric acid and chlorovinyl arsenous oxide, a vesicant. The latter is a nonvolatile solid that is not readily washed away by rains. Strong alkalies destroy these blister-forming properties.
Rate of hydrolysis	Rapid for vapor and dissolved Lewisite L-2. Low solubility in water limits the hydrolysis.
Stability	Reasonably stable; however, in presence of moisture, it hydrolyses rapidly, losing its vesicant property.
Storage stability	Stable in steel or glass container, corrosive to steel

Decomposition	It hydrolyses in acidic medium to form HCl and nonvolatile (solid) chlorovinylarsenious oxide, which is a less potent vesicant than Lewisite L-2. Hydrolysis in alkaline medium, as in decontamination with alcoholic caustic or carbonate solution, produces acetylene and trisodium arsenate (Na_3ASO_4). Therefore, decontaminated solution would contain toxic arsenic.
Polymerization	No data available

Toxicity

LD_{50} (skin)	8 mg/kg
LCt_{50} (respiratory)	No data available
LCt_{50} (percutaneous)	No data available
ICt_{50} (respiratory)	No data available
Ict_{50} (percutaneous)	No data available
Rate of detoxification	The body does not detoxify Lewisite L-2
Skin and eye toxicity	Even limited concentrations of L-2 vapor (below 300 mg-min/m^3) cause extreme irritation of the eyes. Burning, pain, sensitivity to light, tearing, and swelling of the eyes result. An exposure of 1,500 mg-min/m^3 produces severe and probably permanent corneal damage to the eyes. Liquids cause severe damage to the eyes. L-2 has about the same blistering action on the skin as HD, even though the lethal dosage for L-2 is much higher.
Rate of action	Rapid. The body absorbs L-2 more rapidly through the skin than the nitrogen mustards.
Overexposure effects	L-2 is a vesicant (blister agent); also, it acts as a systemic poison, causing pulmonary edema, diarrhea, restlessness, weakness, subnormal temperature, and low blood pressure. In order of severity and appearance of symptoms, it is a blister agent, a toxic lung irritant, absorbed in tissues, and a systemic poison. When inhaled in high concentrations, it may be fatal in as short a time as 10 min. L-2 is not detoxified by the body. Common routes of entry into the body include ocular, percutaneous, and inhalation.

Safety

Protective gloves	Wear Norton Chemical Protection Glove Set, M3 butyl rubber
Eye protection	Wear protective eyeglasses as a minimum; use goggle sand face shield for splash hazards.
Other	Wear full protective clothing (Level A) consisting of M3 butyl rubber suit with hood, M2A1 boots, M3 gloves,

impregnated underwear, M9 series mask and coveralls (if desired), or the 30 mil DPE if available, or NIOSH-approved equivalent; wear gloves and lab coat with M9, M17, or M40 mask readily available for general lab work. In addition, wear daily clean smock, foot covers, and head covers when handling contaminated lab animals.

Emergency procedures

Inhalation: Remove from the source immediately; give artificial respiration if breathing has stopped; administer oxygen if breathing is difficult; seek medical attention immediately.

Eye Contact: Speed in decontaminating the eyes is absolutely essential; remove person from the liquid source; flush the eyes immediately with water for 10–15 min by tilting the head to the side, pulling eyelids apart with fingers, and pouring water slowly into the eyes; do not cover eyes with bandages, but if necessary, protect eyes by means of dark or opaque goggles; seek medical attention immediately.

Skin Contact: Remove victim from source immediately and remove contaminated clothing; immediately decon affected areas by flushing with 10% sodium carbonate solution; wash off with soap and water after 3–4 min to protect against erythema; seek medical attention immediately.

Ingestion: Do not induce vomiting; give victim milk to drink; seek medical attention immediately.

Military Significant Information

Field protection

Protective mask and permeable protective clothing for L-2 vapor and small droplets; impermeable protective clothing for protection against large droplets, splashes, and smears.

Decontamination

HTH, STB, household bleach, DS2, or caustic soda. Decontaminate liquid agent on the skin with the M258A1, M258, or M291 skin decontaminating kit. Decontaminate individual equipment with the M280 individual equipment decontamination kit.

Persistency

Somewhat shorter than for L-1; very short duration under humid conditions.

Use

Moderately delayed-action casualty agent. Not authorized for military use.

1.5. Lewisite – L-3

L-3 is a vesicant (blister agent); also, it acts as a systemic poison, causing pulmonary edema, diarrhea, restlessness, weakness, subnormal temperature, and low blood pressure. In order of severity and appearance of symptoms, it is: a blister agent, a toxic lung irritant, absorbed in tissues, and a systemic poison. When inhaled in high concentrations, it may be fatal in as short a time as 10 min. L-3 is not detoxified by the body. Common routes of entry into the body include ocular, percutaneous, and inhalation.

Informational

Designation	L-3
Class	Blister agent
Type	B – persistent
Chemical name	Tris (2-chloroethenyl) Arsine
CAS number	[40334-70-1]

Chemical and Physical Properties

Appearance	Colorless to brownish
Odor	Fruity or flowery odor
Chemical formula	$C_6H_6AsCl_3$
Molecular weight	259.39
Chemical structure	$$ClCH = CH - As - CH = CHCl$$ with $CH=CHCl$ on the As
Melting point	18–23°C
Boiling point	260°C
Flash point	No data available
Decomposition temperature	No data available
Vapor density	No data available
Liquid density	1.57 g/cm^3 at 20°C
Solid density	No data available
Vapor pressure	No data available
Volatility	No data available
Solubility	No data available

Reactivity

Hydrolysis products	Hydrochloric acid and chlorovinyl arsenous oxide, a vesicant. The latter is a nonvolatile solid that is not readily washed away by rains. Strong alkalies destroy these blister-forming properties.
Rate of hydrolysis	Rapid for vapor and dissolved Lewisite L-3. Low solubility in water limits the hydrolysis.
Stability	Reasonably stable; however, in presence of moisture, it hydrolyses rapidly, losing its vesicant property.
Storage stability	Stable in steel or glass container. Corrosive to steel.

Decomposition	It hydrolyses in acidic medium to form HCl and nonvolatile (solid) chlorovinylarsenious oxide, which is a less potent vesicant than Lewisite L-3. Hydrolysis in alkaline medium, as in decontamination with alcoholic caustic or carbonate solution, produces acetylene and trisodium arsenate (Na_3AsO_4). Therefore, decontaminated solution would contain toxic arsenic.
Polymerization	No data available

Toxicity

LD_{50} *(skin)*	No data available
LCt_{50} *(respiratory)*	No data available
LCt_{50} *(percutaneous)*	No data available
ICt_{50} *(respiratory)*	No data available
ICt_{50} *(percutaneous)*	No data available
Rate of detoxification	The body does not detoxify Lewisite L-3
Skin and eye toxicity	Even limited concentrations of L-3 vapor (below 300 mg-min/m³) cause extreme irritation of the eyes. Burning, pain, sensitivity to light, tearing, and swelling of the eyes result. An exposure of 1,500 mg-min/m³ produces severe and probably permanent corneal damage to the eyes. Liquids cause severe damage to the eyes. L-3 has about the same blistering action on the skin as HD, even though the lethal dosage for L-3 is much higher.
Rate of action	Rapid. The body absorbs L-3 more rapidly through the skin than the nitrogen mustards.
Overexposure effects	L-3 is a vesicant (blister agent); also, it acts as a systemic poison, causing pulmonary edema, diarrhea, restlessness, weakness, subnormal temperature, and low blood pressure. In order of severity and appearance of symptoms, it is a blister agent, a toxic lung irritant, absorbed in tissues, and a systemic poison. When inhaled in high concentrations, it may be fatal in as short a time as 10 min. L-3 is not detoxified by the body. Common routes of entry into the body include ocular, percutaneous, and inhalation.

Safety

Protective gloves	Wear Norton Chemical Protection Glove Set, M3 butyl rubber.
Eye protection	Wear protective eyeglasses as a minimum; use goggle sand face shield for splash hazards.
Other	Wear full protective clothing (Level A) consisting of M3 butyl rubber suit with hood, M2A1 boots, M3 gloves, impregnated underwear, M9 series mask and coveralls (if desired), or the 30 mil DPE if available, or NIOSH-approved equivalent; wear

gloves and lab coat with M9, M17, or M40 mask readily available for general lab work. In addition, wear daily clean smock, foot covers, and head covers when handling contaminated lab animals.

Emergency procedures

Inhalation: Remove from the source immediately; give artificial respiration if breathing has stopped; administer oxygen if breathing is difficult; seek medical attention immediately.

Eye Contact: Speed in decontaminating the eyes is absolutely essential; remove person from the liquid source; flush the eyes immediately with water for 10–15 min by tilting the head to the side, pulling eyelids apart with fingers, and pouring water slowly into the eyes; do not cover eyes with bandages, but if necessary, protect eyes by means of dark or opaque goggles; seek medical attention immediately.

Skin Contact: Remove victim from source immediately and remove contaminated clothing; immediately decon affected areas by flushing with 10% sodium carbonate solution; wash off with soap and water after 3–4 min to protect against erythema; seek medical attention immediately.

Ingestion: Do not induce vomiting; give victim milk to drink; seek medical attention immediately.

Military Significant Information

Field protection

Protective mask and permeable protective clothing for L-3 vapor and small droplets; impermeable protective clothing for protection against large droplets, splashes, and smears.

Decontamination

HTH, STB, household bleach, DS2, or caustic soda. Decontaminate liquid agent on the skin with the M258A1, M258, or M291 skin decontaminating kit. Decontaminate individual equipment with the M280 individual equipment decontamination kit.

Persistency

Somewhat shorter than for HD; very short duration under humid conditions.

Use

Moderately delayed-action casualty agent. Not authorized for military use.

1.6. Methyldichloroarsine – MD

MD is similar to ED. Like L, MD causes immediate irritation of the eyes and nose with blistering effects delayed for hours. MD is irritating to the respiratory tract and produces lung injury upon sufficient exposure. The vapor irritates the eyes, and the liquid may severely injure the eyes. The absorption of either vapor or liquid through the skin in sufficient amounts may lead to systemic poisoning or death. Prolonged contact with either liquid or vapor produces blistering of the skin. Vapor concentrations required for blistering effect are very difficult to attain in the field.

Informational

Designation	MD
Class	Blister agent
Type	A – nonpersistent
Chemical name	Methyldichloroarsine
CAS number	[593-89-5]

Chemical and Physical Properties

Appearance	Colorless liquid
Odor	None
Chemical formula	CH_3AsCl_2
Molecular weight	160.86
Chemical structure	$CH_3 — As \diagup^{Cl} \diagdown_{Cl}$
Melting point	–55°C
Boiling point	133°C
Flash point	No data available
Decomposition temperature	Stable to boiling point
Vapor density	5.5 (air = 1)
Liquid density	1.836 g/cm^3 at 20°C
Solid density	No data available
Vapor pressure	2.17 mmHg at 0°C 7.76 mmHg at 20°C
Volatility	74,900 mg-min/m^3 at 20°C
Solubility	Soluble in ethyl chloride, alcohol, ether, benzene, acetone, and cyclohexane

Reactivity

Hydrolysis products	Hydrochloric acid and methylarsenic oxide
Rate of hydrolysis	Very rapid
Stability	No data available
Storage stability	Stable in steel containers
Decomposition	No data available
Polymerization	No data available

Toxicity

LD$_{50}$ (skin)	No data available
LCt$_{50}$ (respiratory)	3,000–5,000 mg-min/m^3
LCt$_{50}$ (percutaneous)	No data available
ICt$_{50}$ (respiratory)	25 mg-min/m^3
ICt$_{50}$ (percutaneous)	No data available
Rate of detoxification	Detoxified at an appreciable rate
Skin and eye toxicity	Blistering action slightly less than that of HD. Has effect on eyes similar to that of L (produces corneal damage) but less severe. Vapor concentration required for blistering effect is very difficult to attain in the field.
Rate of action	Immediate irritation of eyes and nose. Blistering effect is delayed several hours.
Overexposure effects	No data available

Safety

Protective gloves	Wear butyl toxicological agent protective gloves (M3, M4, glove set).
Eye protection	Wear chemical goggles as a minimum; use goggles and face shield for splash hazard.
Other	Wear gloves and lab coat with M9 or M17 mask readily available for general lab work. In addition, wear daily clean smock, foot covers, and head cover when handling contaminated lab animals.
Emergency procedures	**Inhalation:** Remove victim from the source immediately; administer artificial respiration if breathing has stopped; administer oxygen if breathing is difficult; seek medical attention immediately.
	Eye Contact: Speed in decontaminating the eyes is absolutely essential; remove person from the liquid source, flush the eyes immediately with water by tilting the head to the side, pulling the eyelids apart with the fingers, and pouring water slowly into the eyes; do not cover eyes with bandages; but if necessary, protect eyes by means of dark or opaque goggles; seek medical attention immediately.
	Skin Contact: Don respiratory protective masks and gloves; remove victim from agent source immediately; flush skin and clothes with 5% solution of sodium hypochlorite or liquid household bleach within 1 min; cut and remove contaminated clothing; flush contaminated skin area again with 5% sodium hypochlorite solution; then wash contaminated skin area with soap and water; seek medical attention immediately.
	Ingestion: Do not induce vomiting; give victim milk to drink; seek medical attention immediately.

Military Significant Information

Field protection   Protective mask and protective clothing.

Decontamination   HTH, STB, household bleach, DS2, or caustic soda. Decontaminate liquid agent on the skin with the M258A1, M258, or M291 skin decontaminating kit. Decontaminate individual equipment with the M280 individual equipment decontamination kit.

Persistency   Relatively short.

Use   Delayed-action casualty agent.

1.7. Mustard – T

T is bis-(2-chloroethylthio)ethyl ether. It is a byproduct of certain methods of mustard manufacture, and is intentionally included in certain munitions to depress the freezing point of the mustard. T is also a vesicant; the toxicology and the chemistry of T are quite similar to the corresponding properties of mustard. Mustard T accounts for 40% of HT mixture.

Informational

Designation   T

Class   Blister agent

Type   B – persistent

Chemical name   Bis-[2-(2-chloroethylthio)ethyl]ether

CAS number   [63918-89-8]

Chemical and Physical Properties

Appearance   T is a sulfur and chlorine compound similar in structure to HD and is a clear yellowish liquid.

Odor   Slight garlic- or mustard-like odor

Chemical formula $C_8H_{16}Cl_2OS_2$

Molecular weight   263.2

Chemical structure $ClCH_2CH_2SCH_2CH_2OCH_2CH_2SCH_2CH_2Cl$

Melting point   9.4°C

Boiling point   373.8°C

Flash point   179.9°C

Decomposition temperature   No data available

Vapor density	9.1 (air = 1)
Liquid density	1.24 at 25°C
Solid density	No data available
Vapor pressure	0.00003 mmHg (25°C)
Volatility	0.42 mg/m^3
Solubility	Insoluble in H_2O

Reactivity

Hydrolysis products	Hydrochloric acid and thioglycols or thioethers may be produced.
Rate of hydrolysis	No data available
Stability	Unstable with oxidizers and will ignite
Storage stability	No data available
Decomposition	Hydrochloric acid and sulfur oxides
Polymerization	Will not occur

Toxicity

LD_{50} (skin)	No data available
LCt_{50} (respiratory)	No data available
LCt_{50} (percutaneous)	No data available
ICt_{50} (respiratory)	No data available
ICt_{50} (percutaneous)	No data available
Rate of detoxification	Slowly detoxified by the body, exposures are essentially cumulative.
Skin and eye toxicity	No data available
Rate of action	Rapid
Overexposure effects	No data available

Safety

Protective gloves	Mandatory; wear butyl toxicological agent protective gloves (M3, M4, gloveset).
Eye protection	Wear chemical goggles as a minimum; use goggles and face shield for splash hazard.
Other	Wear gloves and lab coat with M9 or M17 mask readily available for general lab work. In addition, wear daily clean

smock, foot covers, and head cover when handling contaminated lab animals.

Emergency
procedures

Inhalation: Remove victim from the source immediately; administer artificial respiration if breathing has stopped; administer oxygen if breathing is difficult; seek medical attention immediately.

Eye Contact: Speed in decontaminating the eyes is absolutely essential; remove victim from the liquid source, flush the eyes immediately with water by tilting the head to the side, pulling the eyelids apart with the fingers, and pouring water slowly into the eyes; do not cover eyes with bandages; but if necessary, protect eyes by means of dark or opaque goggles; seek medical attention immediately.

Skin Contact: Don respiratory protective masks and gloves; remove victim from agent source immediately; flush skin and clothes with 5% solution of sodium hypochlorite or liquid household bleach within 1 min; cut and remove contaminated clothing; flush contaminated skin area again with 5% sodium hypochlorite solution; then wash contaminated skin area with soap and water; seek medical attention immediately.

Ingestion: Do not induce vomiting; give victim milk to drink; seek medical attention immediately.

Military Significant Information

Field protection

Protective mask and permeable protective clothing for vapor and small droplets; impermeable protective clothing for protection against large droplets, splashes, and smears.

Decontamination

HTH, STB, household bleach, fire or DS2. Decontaminate liquid agent on the skin with the M258A1, M258, or M291 skin decontaminating kit. Decontaminate individual equipment with the M280 individual equipment decontamination kit.

Persistency

Heavily splashed T liquid persists 1–2 days under average weather conditions, and a week or more under very cold conditions.

Use

Delayed-action casualty agent.

1.8. Mustard – Lewisite Mixture – HL

HL is a lethal vesicant and alkylating agent producing cytotoxic action on the hematopoietic (blood-forming) tissues and is not detoxified in the body. Locally, HL affects both the skin and eyes. HL is a liquid mixture of mustard (HD) and Lewisite (L) designed to provide a low freezing point for use in cold weather and high altitudes. The eutectic mixture (lowest freezing point) is 63% lewisite and 37% mustard. HL has a garlic-like odor from its HD content.

Informational

Designation	HL
Class	Blister agent
Type	B – persistent
Chemical name	HD: bis-(2-chloroethyl) sulfide L1: 2-Chlorovinyldichloroarsine
CAS number	Mixture: [Not available]; HD: [505-60-2]; L1: [541-25-3]

Chemical and Physical Properties

Appearance	Yellow to brown oily liquid
Odor	Garlic-like
Chemical formula	HD: $C_4H_8Cl_2S$; L1: $C_2H_2AsCl_3$
Molecular weight	HD: 159.08; L1: 207.32; 186.4, based on the eutectic mixture (37% HD and 63% L1)

$$Cl - CH_2CH_2 - S - CH_2CH_2 - Cl$$

+

Chemical structure

$$ClCH = CH - As \Big\backslash^{Cl}_{Cl}$$

Melting point	–25.4°C when pure –42°C plant purity (calculated)
Boiling point	<190°C decomposes before boiling
Flash point	No data available
Decomposition temperature	>100°C
Vapor density	6.4 (air = 1)
Liquid density	1.66 g/cm^3 at 20°C
Solid density	No data available
Vapor pressure	0.02 mmHg at –10°C 0.248 mmHg at 20°C 1.03 mmHg at 40°C
Volatility	240 mg/m^3 at –11°C 2,730 mg/m^3 at 20°C 10,270 mg/m^3 at 40°C (Calculated from above vapor pressure; actual volatility is somewhat lower.)
Solubility	Practically insoluble in water.

Reactivity

Hydrolysis products   Hydrochloric acid, thiodiglycol, and chlorovinylarsenious oxide. Alkaline hydrolysis destroys the blistering properties.

Rate of hydrolysis   Rapid in the liquid or vapor state; slow at ordinary temperatures.

Stability   HL is a persistent agent depending on pH and moisture and has been known to remain active for up to 3 years in soil.

Storage stability   Stable at ambient temperatures and in lacquered steel containers. Conditions to avoid: rapidly corrosive to brass at 65˚C; will corrode steel at a rate of 0.0001 in. of steel per month at 65˚C.

Decomposition   HL will hydrolyze into hydrochloric acid, thiodiglycol, and nonvesicant arsenic compounds.

Polymerization   Will not occur.

Toxicity

LD_{50} (skin)   No data available

LCt_{50} (respiratory)   >1,500 mg-min/m^3

LCt_{50} (percutaneous)   >10,000 mg-min/m^3

ICt_{50} (respiratory)   No data available

ICt_{50} (percutaneous)   1,500–2,000 mg-min/m^3

Rate of detoxification   Not detoxified

Skin and eye toxicity   Very high

Rate of action   Produces immediate stinging of skin and redness within 30 min; blistering delayed about 13 h.

Overexposure effects   HL is a vesicant and alkylating agent producing cytotoxic action on the hematopoietic (blood-forming) tissues, which are especially sensitive. The rate of detoxification of HL in the body is very slow, and repeated exposure produces a cumulative effect. Contamination of the skin produces immediate stinging of the skin, turning red within 30 min. Blistering is delayed for about 13 h and tends to cover the entire area of reddened skin. Blisters from HL exposures are deeper and more painful than with HD. Local action on the eyes is extremely rapid, and produces severe necrotic damage and loss of eyesight. Exposure of eyes to HL vapor or aerosol produces lacrimation, photophobia, and inflammation of the conjunctiva and cornea. When HL vapor/aerosol is inhaled, the respiratory tract becomes inflamed after a few hours latency period, accompanied by sneezing, coughing, and bronchitis, diarrhea, and fever. The respiratory damage is similar to that produced by mustard, except in the most severe cases. In these cases, fluid in the chest cavity may accompany fluid in the lungs. HL is absorbed through skin contact and inhalation of vapors, causing systemic toxicity such as damage to the lungs, bone marrow, lymph nodes, spleen, and endocrine system.

Safety

Protective gloves	Wear butyl toxicological agent protective gloves (M3, M4, or glove set).
Eye protection	Wear chemical goggles as a minimum; use goggles and face shield for splash hazard.
Other	Wear gloves and lab coat for general lab work; have an M9, M40, or M17 mask readily available.
Emergency procedures	*Inhalation:* Remove from the source immediately; give artificial respiration if breathing has stopped; administer oxygen if breathing is difficult; seek medical attention immediately.
	Eye Contact: Speed in decontaminating the eyes are absolutely essential; remove person from the liquid source; flush the eyes immediately with water by tilting the head to the side, pulling the eyelids apart with the fingers and pouring water slowly into the eyes; do not cover eyes with bandages; but if necessary, protect eyes by means of dark of opaque goggles; seek medical attention immediately.
	Skin Contact: Don respiratory protective mask and gloves; remove victim from agent source immediately; flush skin and clothes with 5% solution of sodium hypochlorite or liquid household bleach within 1 min; cut and remove contaminated clothing; flush contaminated skin area again with 5% sodium hypochlorite solution, then wash contaminated skin area with soap and water; wash thoroughly if shower facilities are available; seek medical attention immediately.
	Ingestion: Do not induce vomiting; give victim milk to drink; seek medical attention immediately.

Military Significant Information

Field protection	Protective mask and permeable protective clothing for vapor and small droplets; impermeable protective clothing for protection against large droplets, splashes, and smears.
Decontamination	Bleach, fire, DS2, or caustic soda. Decontaminate liquid agent on the skin with the M258A1, M258, or M291 skin decontaminating kit. Decontaminate individual equipment with the M280 individual equipment decontamination kit.

Persistency	Depends on munitions used and the weather. Somewhat shorter than that of HD, heavily splashed liquid of which persists 1–2 days under average weather conditions, and a week or more under very cold conditions.
Use	Delayed-action casualty agent.

1.9. Nitrogen Mustard – HN-1

HN-1 is a vesicant and an alkylating agent, producing cytotoxic action on the hematopoietic (blood-forming) tissues. HN-1 was the first compound of the HN series developed in the late 1920s and early 1930s. HN-1 was designed as a pharmaceutical (wart remover) and became a military agent; HN-2 was designed as a military agent and became a pharmaceutical; HN-3 was designed as a military agent and is the only one of these agents that remains anywhere as a military agent. These agents are more immediately toxic than the sulfur mustards.

Informational

Designation	HN-1
Class	Blister agent
Type	B – persistent
Chemical name	Bis-(2-chloroethyl)ethylamine
CAS number	[538-07-8]

Chemical and Physical Properties

Appearance	HN-1 is oily, colorless to pale yellow
Odor	A faint, fishy, or musty odor
Chemical formula	$C_6H_{13}Cl_2N$
Molecular weight	170.08
Chemical structure	$CH_3CH_2-N \begin{smallmatrix} \diagup CH_2CH_2Cl \\ \diagdown CH_2CH_2Cl \end{smallmatrix}$
Melting point	–34°C
Boiling point	194°C (calculated); decomposes. At atmospheric pressure HN-1 decomposes below boiling point.
Flash point	No immediate danger of fire or explosion.
Decomposition temperature	Decomposes before boiling point is reached.
Vapor density	5.9 (air = 1)

Liquid density	1.09 g/cm^3 at 25°C
Solid density	No data available
Vapor pressure	0.0773 at 10°C
	0.25 at 25°C
	0.744 at 40°C
Volatility	127 mg/m^3 at –10°C
	308 mg/m^3 at 0°C
	1,520 mg/m^3 at 20°C
	3,100 mg/m^3 at 30°C
Solubility	Sparingly soluble in water; freely soluble in acetone and other organic solvents.

Reactivity

Hydrolysis products	Hydroxyl derivatives and condensation products (all intermediate hydrolysis products are toxic).
Rate of hydrolysis	Slow because of low solubility in water; less readily hydrolyzed than mustard.
Stability	Polymerizes slowly
Storage stability	Adequate for use in munitions. Polymerizes slowly. Corrosive to ferrous alloys beginning at 65°C.
Decomposition	Toxic intermediate products are produced during hydrolysis. Approximate half-life in water at 25°C is 1.3 min. Decomposition comes through slow change into quaternary ammonium salts. Decomposition point is below 94°C.
Polymerization	Polymerizes slowly.

Toxicity

LD$_{50}$ (skin)	No data available
LCt$_{50}$ (respiratory)	1,500 mg-min/m^3
LCt$_{50}$ (percutaneous)	20,000 mg-min/m^3
ICt$_{50}$ (respiratory)	No data available
ICt$_{50}$ (percutaneous)	9,000 mg-min/m^3
Rate of detoxification	Not detoxified Cumulative.
Skin and eye toxicity	Eyes are very susceptible to low concentrations, incapacitating effects by skin absorption require higher concentrations.
Rate of action	Delayed; 12 h or longer.
Overexposure effects	The vapors are irritating to the eyes and nasal membranes even in low concentration. HN-1 is a vesicant (blister agent)

and alkylating agent producing cytotoxic action on the hematopoietic (blood-forming) tissues. HN-1 is not naturally detoxified by the body; therefore, repeated exposure produces a cumulative effect.

Safety

Protective gloves   Mandatory; wear butyl toxicological agent protective gloves (M3, M4, or glove set).

Eye protection   Wear chemical goggles as a minimum; use goggles and face shield for splash hazard.

Other   Wear full protective clothing consisting of the M3 butyl rubber suit with hood, M2A1 boots, M3 gloves, treated underwear, M9 series mask and coveralls (if desired). Wear gloves and lab coat with M9, M17, or M40 Mask readily available for general lab work. In addition, wear daily clean smock, foot covers, and head cover when handling contaminated lab animals.

Emergency procedures   **Inhalation:** Remove from source immediately; give artificial respiration if breathing has stopped; administer oxygen if breathing is difficult; seek medical attention immediately.

Eye Contact: Flush eyes immediately with water for 10–15 min, pulling eyelids apart with fingers, and pouring water into eyes; do not cover eyes with bandages; protect eyes with dark or opaque goggles after flushing eyes; seek medical attention immediately.

Skin Contact: Don respiratory mask and gloves; remove victim from source immediately and remove contaminated clothing; decontaminate the skin immediately by flushing with a 5% solution of liquid household bleach; wash off with soap and water after 3–4 min to remove decon agent and protect against erythema; seek medical attention immediately; to prevent systemic toxicity, decontamination should be done as late as 2 or 3 h after exposure even if it increases the severity of the local reaction; further cleans with soap and water.

Ingestion: Do not induce vomiting; give victims milk to drink; seek medical attention immediately.

Military Significant Information

Field protection   Protective mask and permeable protective clothing for vapor and small droplets; impermeable protective clothing for protection against large droplets, splashes, and smears.

Decontamination   HTH, STB, household bleach, fire or DS2. Decontaminate liquid agent on the skin with the M258A1, M258, or M291 skin decontaminating kit. Decontaminate individual equipment with the M280 individual equipment decontamination kit.

Persistency   Depends on munitions used and the weather; somewhat shorter duration of effectiveness for HD, heavily splashed liquid of which persists 1–2 days under average weather conditions, and a week or more under very cold conditions.

Use   Delayed-action casualty agent.

1.10. Nitrogen Mustard – HN-2

HN-2, the second of a series of nitrogen mustard compounds developed in the late 1920s and early 1930s, was designed as a military agent which became a pharmaceutical substance called Mustine. The chemical intermediate it produces is used as an antineoplastic drug. These agents are more immediately toxic than the sulfur mustards. HN-2 is highly unstable and is no longer seriously considered as a chemical agent. It is rated as somewhat more toxic than HN-1.

Informational

Designation   HN-2

Class   Blister agent

Type   B – persistent

Chemical name   Bis-(2-chloroethyl)methylamine

CAS number   [51-75-2]

Chemical and Physical Properties

Appearance   HN-2 is pale amber to yellow oily liquid.

Odor   Fruity odor in high concentrations; smells like soft soap with a fishy smell in low concentrations.

Chemical formula $C_5H_{11}Cl_2N$

Molecular weight   156.07

Chemical structure $CH_3-N \begin{cases} CH_2CH_2Cl \\ CH_2CH_2Cl \end{cases}$

Melting point   –65°C to –60°C

Boiling point   75°C at 15 mmHg; at atmospheric pressure HN-2 decomposes below boiling point.

Flash point   No immediate danger of fire or explosion.

Decomposition temperature	Decomposes before boiling point is reached. Instability of HN-2 is associated with its tendency to polymerize or condense; the reactions involved could generate enough heat to cause an explosion.
Vapor density	5.4 (air = 1)
Liquid density	1.15 g/cm^3 at 20°C
Solid density	No data available
Vapor pressure	0.130 at 10°C 0.290 at 20°C 0.427 at 25°C 1.25 at 40°C
Volatility	1,150 mg/m^3 at 10°C 3,580 mg/m^3 at 25°C 5,100 mg/m^3 at 30°C 10,000 mg/m^3 at 40°C
Solubility	Soluble in acetone and organic solvents and oil; sparingly soluble in water.

Reactivity

Hydrolysis products	Complex condensates or polymers
Rate of hydrolysis	Slow, except in presence of alkalis; products formed are complex polymeric quaternary ammonium salts; dimerizes fairly rapidly in water.
Stability	Not stable; decomposes before boiling point is reached or condenses under all conditions; the reactions involved could generate enough heat to cause an explosion; dry crystals are stable.
Storage stability	Not stable
Decomposition	Approximate half-life in water at 25°C is 4 min; decomposition point is below boiling point.
Polymerization	Polymerized components will present an explosion hazard in open air.

Toxicity

LD_{50} *(skin)*	No data available
LCt_{50} *(respiratory)*	3,000 $mg\text{-}min/m^3$
LCt_{50} *(percutaneous)*	No data available
ICt_{50} *(respiratory)*	No data available
ICt_{50} *(percutaneous)*	6,000–9,500 $mg\text{-}min/m^3$

Rate of detoxification	Not detoxified.
Skin and eye toxicity	HN-2 has the greatest blistering power of the nitrogen mustards in vapor form but is intermediate as a liquid blistering agent. It produces toxic eye effects more rapidly than does HD.
Rate of action	Skin effects delayed 12 h or longer.
Overexposure effects	HN-2 is highly irritating to the eyes and throat; in highconcentrations it can cause blindness. Absorbed into the bloodstream it will seriously interfere with the functioningof hemoglobin and will eventually damage the endocrinesystem. HN-2 is a vesicant (blister agent) and alkylating agent producing cytotoxic action on the hematopoietic (blood-forming) tissues which are especially sensitive. HN-2 is not naturally detoxified by the body; therefore, repeated exposure produces a cumulative effect.

Safety

Protective gloves	Mandatory; wear butyl toxicological agent protective gloves (M3, M4, or glove set).
Eye protection	Wear chemical goggles as a minimum; use goggles and face shield for splash hazard.
Other	Wear full protective clothing consisting of the M3 butyl rubber suit with hood, M2A1 boots, M3 gloves, treated underwear, M9 series mask and coveralls (if desired). Wear gloves and lab coat with M9, M17, or M40 mask readily available for general lab work.
Emergency procedures	***Inhalation:*** Remove from source immediately; give artificial respiration if breathing has stopped; administer oxygen if breathing is difficult; seek medical attention immediately.
	Eye Contact: Flush eyes immediately with water for 10–15 min, pulling eyelids apart with fingers, and pouring water into eyes; do not cover eyes with bandages; protect eyes with dark or opaque goggles after flushing eyes; seek medical attention immediately.
	Skin Contact: Don respiratory mask and gloves; remove victim from source immediately and remove contaminated clothing; decontaminate the skin immediately by flushing with 5% solution of liquid household bleach; wash off with soap and water after 3–4 min to remove decon agent and protect against erythema; seek medical attention immediately; to prevent systemic toxicity, decontaminate as late as 2 or 3 h after exposure even if it increases the severity of the local reaction; further clean with soap and water.

Ingestion: Do not induce vomiting; give victims milk to drink; seek medical attention immediately.

Military Significant Information

Field protection	Protective mask and permeable protective clothing for ED vapor and small droplets; impermeable protective clothing for protection against large droplets, splashes, and smears.
Decontamination	STB, fire, or DS2. Decontaminate liquid agent on the skin with the M258A1, M258, or M291 skin decontaminating kit. Decontaminate individual equipment with the M280 individual equipment decontamination kit.
Persistency	Depends on munitions used and the weather; somewhat shorter duration of effectiveness for HD, heavily splashed liquid of which persists 1–2 days under average weather conditions, and a week or more under very cold conditions.
Use	Delayed-action casualty agent.

1.11. Nitrogen Mustard – HN-3

HN-3 was the last of the nitrogen mustard agents developed. It was designed as a military agent and is the only one of the nitrogen mustards that is still used for military purposes. It is the principal representative of the nitrogen mustards because its vesicant properties are almost equal to those of HD. It also is the most stable in storage of the three nitrogen mustards.

Informational

Designation	HN-3
Class	Blister agent
Type	B – persistent
Chemical name	2, 2', 2"-trichlorotriethylamine
CAS number	[555-77-1]

Chemical and Physical Properties

Appearance	HN-3 is a colorless to pale yellow liquid. It is also the most stable in storage of three nitrogen mustards.
Odor	Butter almond odor

Chemical formula	$C_6H_{12}Cl_3N$
Molecular weight	204.54

Chemical structure

$$ClCH_2CH_2 \;-\; N \overset{\displaystyle CH_2CH_2Cl}{\underset{\displaystyle CH_2CH_2Cl}{\Big\langle}}$$

Melting point	$-3.7°C$
Boiling point	$256°C$ (calculated) decomposes. At atmospheric pressure HN-2 decomposes below boiling point.
Flash point	No data available
Decomposition temperature	Decomposes before boiling point is reached.
Vapor density	7.1 (air = 1)
Liquid density	1.24 g/cm^3 at 25°C
Solid density	No data available
Vapor pressure	0.0109 mmHg at 25°C
Volatility	13 mg/m^3 at 0°C 121 mg/m^3 at 25°C 180 mg/m^3 at 30°C 390 mg/m^3 at 40°C
Solubility	Soluble in sulfur mustards and chloropicrin; insoluble in water; soluble in ether, benzene, and acetone.

Reactivity

Hydrolysis products	Hydrochloric acid and triethanolamine in dilute solutions. Dimer formation in higher concentrations.
Rate of hydrolysis	Slow, except in presence of alkalis; products formed are complex polymeric quaternary ammonium salts; dimerizes fairly rapidly in water.
Stability	Slow but steady polymerization; not stable; decomposes before boiling point is reached or condenses under all conditions; the reactions involved could generate enough heat to cause an explosion.
Storage stability	Stable enough for use as a bomb filling even under tropical conditions. However, the agent darkens and deposits a crystalline solid in storage.
Decomposition	Approximate half-life in water at 25°C is 4 min; decomposition point is below boiling point.
Polymerization	Polymerized components will present an explosion hazard in open air.

Toxicity

LD$_{50}$ (skin)	0.7 gm/person (estimated)
LCt$_{50}$ (respiratory)	1,500 mg-min/m^3
LCt$_{50}$ (percutaneous)	10,000 mg-min/m^3
ICt$_{50}$ (respiratory)	No data available
ICt$_{50}$ (percutaneous)	2,500 mg-min/m^3. This information is based on estimates and indicates that HN-3 closely approaches HD in vapor toxicity and that it is the most toxic of the nitrogen mustards.
Rate of detoxification	Not detoxified; cumulative
Skin and eye toxicity	No data available
Rate of action	Most symptoms are delayed 4–6 h (same as for HD). In some cases eye irritation, tearing, and sensitivity to light (photophobia) develop immediately.
Overexposure effects	HN-3 is a cumulative poison which is highly irritating to the eyes and throat. Eye irritation, tearing, and photophobia develop immediately after exposure. The median incapacitating dose for eyes is 200 mg-min/m^3. Blistering of the skin may occur after liquid exposure, severe or persistent exposure, or vapor condensation in sweat. Usually a rash will develop from liquid contamination within an hour, replaced by blistering between 6 and 12 h after exposure. HN-3 interferes with hemoglobin functioning in the blood, hindering the production of new blood cells and destroying white blood.

Safety

Protective gloves	Mandatory; wear butyl toxicological agent protective gloves (M3, M4, or glove set).
Eye protection	Wear chemical goggles as a minimum; use goggles and face shield for splash hazard.
Other	Wear full protective clothing consisting of the M3 butyl rubber suit with hood, M2A1 boots, M3 gloves, underwear, M9 series mask and coveralls (if desired). For general lab work, wear gloves and lab coat with M9, M17, or M40 mask readily available. In addition, wear daily clean smock, foot covers, and head cover when handling contaminated lab animals.
Emergency procedures	**Inhalation:** Remove from source immediately; give artificial respiration if breathing has stopped; administer oxygen if breathing is difficult; seek medical attention immediately.
	Eye Contact: Flush eyes immediately with water for 10–15 min, pulling eyelids apart with fingers and pouring water into eyes; do not cover eyes with bandages; protect eyes with dark or opaque goggles after flushing eyes; seek medical attention immediately.

Skin Contact: Don respiratory mask and gloves; remove victim from source immediately and remove contaminated clothing; decontaminate the skin immediately by flushing with a 5% solution of liquid household bleach; wash off with soap and water after 3–4 min to remove decon agent and protect against erythema; seek medical attention immediately; to prevent systemic toxicity, decontaminate as late as 2 or 3 h after exposure even if it increases the severity of the local reaction; further clean with soap and water.

Ingestion: Do not induce vomiting; give victims milk to drink; seek medical attention immediately.

Military Significant Information

Field protection	Protective mask and permeable protective clothing for ED vapor and small droplets; impermeable protective clothing for protection against large droplets, splashes, and smears.
Decontamination	STB, fire, or DS2. Decontaminate liquid agent on the skin with the M258A1, M258, or M291 skin decontaminating kit. Decontaminate individual equipment with the M280 individual equipment decontamination kit.
Persistency	Considerably longer than for HD. HN-3 use is emphasized for terrain denial. It can be approximately 2× or 3× the persistence of HD and adheres well to equipment and personnel especially in cold weather.
Use	Delayed-action casualty agent.

1.12. Phenyldichloroarsine – PD

Although PD is classed as a blister agent, it also acts as a vomiting compound. Limited use of PD during World War I did not indicate any marked superiority over the other vomiting compounds used. PD has an immediate effect on eyes and a delayed effect of 30 min to 1 h on skin. PD blisters bare skin but wet clothing decomposes it immediately. The protective mask and protective clothing provide adequate protection, but protection against large droplets, splashes, and smears requires impermeable clothing.

Informational

Designation	PD
Class	Blister agent
Type	B – persistent
Chemical name	Phenyldichloroarsine
CAS number	[696-28-6]

Chemical and Physical Properties

Appearance	Colorless liquid
Odor	None
Chemical formula	$C_6H_5AsCl_2$
Molecular weight	222.91

Chemical structure

Melting point	–20°C
Boiling point	252–255°C
Flash point	No data available
Decomposition temperature	Stable to boiling point
Vapor density	7.7 (air = 1)
Liquid density	1.65 g/cm³ at 20°C
Solid density	No data available
Vapor pressure	0.033 mmHg at 25°C 0.113 mmHg at 40°C
Volatility	390 mg/m³ at 25°C. If dispersed as an aerosol, it would be effective against unprotected troops although only as an agent with a short duration of effectiveness.
Solubility	Slightly soluble in water; miscible with alcohol, benzene, kerosene, petroleum, and olive oil.

Reactivity

Hydrolysis products	Hydrochloric acid and phenylarsenious oxide
Rate of hydrolysis	Rapid
Stability	No data available
Storage stability	Very stable
Decomposition	No data available
Polymerization	No data available

Toxicity

LD_{50} *(skin)*	No data available
LCt_{50} *(respiratory)*	2,600 mg-min/m³

LCt_{50} *(percutaneous)*	No data available
ICt_{50} *(respiratory)*	No data available
ICt_{50} *(percutaneous)*	16 mg-min/m^3 as a vomiting agent; 0 mg-min/m^3 as a vesicant
Rate of detoxification	No specific information but, like related compounds, PD is probably detoxifies rapidly in sub-lethal dosages.
Skin and eye toxicity	About 30% as toxic to the eyes as HD; 633 mg-min/m^3 would produce casualties by eye injury. On bare skin PD is about 90% as blistering as HD, but wet clothing decomposes it immediately.
Rate of action	Immediate effect on eyes; effects on skin are delayed 30 min to 1 h.
Overexposure effects	No data available

Safety

Protective gloves	Wear butyl toxicological agent protective gloves (M3, M4, glove set).
Eye protection	Wear chemical goggles as a minimum; use goggles and face shield for splash hazard.
Other	Wear gloves and lab coat with M9 or M17 mask readily available for general lab work. In addition, wear daily clean smock, foot covers, and head cover when handling contaminated lab animals.
Emergency procedures	**Inhalation**: Remove victim from the source immediately; administer artificial respiration if breathing has stopped; administer oxygen if breathing is difficult; seek medical attention immediately.
	Eye Contact: Speed in decontaminating the eyes is absolutely essential; remove person from the liquid source, flush the eyes immediately with water by tilting the head to the side, pulling the eyelids apart with the fingers, and pouring water slowly into the eyes; do not cover eyes with bandages; but if necessary, protect eyes by means of dark or opaque goggles; seek medical attention immediately.
	Skin Contact: Don respiratory protective masks and gloves; remove victim from agent source immediately; flush skin and clothes with 5% solution of sodium hypochlorite or liquid household bleach within 1 min; cut and remove contaminated clothing; flush contaminated skin area again with 5% sodium hypochlorite solution; then wash contaminated skin area with soap and water; seek medical attention immediately.
	Ingestion: Do not induce vomiting; give victim milk to drink; seek medical attention immediately.

Military Significant Information

Field protection	Protective mask and permeable protective clothing for vapor and small droplets; impermeable protective clothing for protection against large droplets, splashes, and smears.
Decontamination	HTH, STB, household bleach, DS2, or caustic soda. Decontaminate liquid agent on the skin with the M258A1, M258, or M291 skin decontaminating kit. Decontaminate individual equipment with the M280 individual equipment decontamination kit.
Persistency	Depends on munitions used and the weather. Somewhat shorter that that of HD under dry conditions; short duration when wet. (Heavily splashed liquid HD persists 1–2 days under average weather conditions, and a week or more under very cold conditions.)
Use	Delayed-action casualty agent.

1.13. Phosgene Oxime – CX

CX is an urticant, producing instant, almost intolerable pain and local tissue destruction immediately on contact with skin and mucous membranes. It is toxic through inhalation, skin and eye exposure, and ingestion. Its rate of detoxification in the body is unknown.

Informational

Designation	CX
Class	Blister agent
Type	A – nonpersistent
Chemical name	Dichloroformoxime
CAS number	[1794-86-1]

Chemical and Physical Properties

Appearance	Colorless solid or liquid CX may appear as a colorless, low melting point (below 39°C, crystalline) solid or as a liquid above 39°C. It has a high vapor pressure, slowly decomposes at normal temperatures.
Odor	It has an intense, penetrating disagreeable, violently irritating odor.
Chemical formula	$CHCl_2NO$
Molecular weight	113.94

Chemical structure	$\begin{array}{c} Cl \\ \diagdown \\ \diagup \\ Cl \end{array} C{=}NOH$
Melting point	35–40°C
Boiling point	129°C with decomposition
Flash point	No data available
Decomposition temperature	<128°C
Vapor density	3.9 (air = 1)
Liquid density	No data available
Solid density	No data available
Vapor pressure	11.2 mmHg at 25°C (solid) 13 mmHg at 40°C (liquid)
Volatility	1.8×10^3 mg/m^3 at 20°C 7.6×10^4 mg/m^3 at 40°C
Solubility	Dissolves slowly but completely in water; soluble in organic solvents.

Reactivity

Hydrolysis products	Carbon dioxide, hydrochloric acid, hydroxylamine.
Rate of hydrolysis	Very slow in H_2O at pH 7; 5% decomposition in 6 days at room temperature; reacts violently in alkaline solution.
Stability	Unstable in metal; store in glass or enamel-lined storage vessels.
Storage stability	Extremely unstable in presence of trace metals or other impurities. Traces of iron chloride may cause explosive decomposition. Pure material is stable for only 1 or 2 months. It may be stabilized by nitromethane, chloropicrin, glycine, ethyl acetate, or ether – but only in glass vessels below 20°C. Apparently, it is most stable in aromatic solvents.
Decomposition	Half-life; gradually decomposes at reflux (129°C); decomposes on storage above –20°C.
Polymerization	Will not occur

Toxicity

LD_{50} *(skin)*	No data available
LCt_{50} *(respiratory)*	No data available
LCt_{50} *(percutaneous)*	3,200 mg-min/m^3 (estimated)
ICt_{50} *(respiratory)*	No data available

ICt$_{50}$ (percutaneous)	Unknown. The lowest irritant concentration after a 10-s exposure is 1 mg/m^3. The effects of the agent become unbearable after 1 min at 3 mg/m^3.
Rate of detoxification	No data available
Skin and eye toxicity	Violently irritating to eyes. Very low concentrations cause copious tearing, inflammation, and temporary blindness. Liquid on skin is corrosive.
Rate of action	Rapid
Overexposure effects	CX vapors are violently irritating to the eyes. Very low concentrations can cause inflammation, lacrimation, and temporary blindness; higher concentrations can cause corneal corrosion and dimming of vision. Contact with the skin can cause skin lesions of the corrosive type. It is characterized by the appearance within 30 s of a central blanched area surrounded by an erythematous ring. Subcutaneous edema follows in about 15 min. After 24 h, the central blanched area becomes necrotic and darkened, and an eschar is formed in a few days. Healing is accompanied by sloughing of the scab; itching may be present throughout healing.

Safety

Protective gloves	Wear butyl toxicological agent protective gloves (M3, M4 or glove set).
Eye protection	Wear chemical goggles as a minimum; use goggles and face shield for splash hazard.
Other	Wear a complete set of protective clothing to include gloves and lab coat for general lab work; have an M9, M40, or M17 mask readily available.
Emergency procedures	**Inhalation:** Remove from the source immediately; give artificial respiration if breathing has stopped; seek medical attention immediately.
	Eye Contact: Flush eyes immediately with copious amounts of water; seek medical attention immediately.
	Skin Contact: Remove victim from the source immediately; decontaminate the skin immediately by flushing with copious amounts of water to remove any phosgene oxime which has not yet reacted with tissue; seek medical attention immediately.
	Ingestion: Do not induce vomiting; seek medical attention immediately.

Military Significant Information

Field protection	A properly fitting protective mask protects the respiratory system. A complete set of protective clothing will protect the remainder of the body.

Decontamination	Use large amounts of water or DS2 on equipment. Because of the rapid reaction of CX with the skin, decontamination will not be entirely effective after pain occurs. Nevertheless, decontaminate as rapidly as possible by flushing the area with large amounts of water to remove any agent that has not reacted with the skin.
Persistency	About 2 h in soil. Relatively nonpersistent on surfaces and water.
Use	Rapid-acting casualty agent.

1.14. Sesquimustard – Q

Sesquimustard has not been typically listed as a chemical warfare agent until it was listed in the CWC Schedule I. It is one of the most powerful vesicants currently known and is highly toxic by inhalation.

Informational

Designation	Q
Class	Blister agent
Type	B – persistent
Chemical name	1,2-bis-(2-chloroethylthio)ethane
CAS number	[3563-36-8]

Chemical and Physical Properties

Appearance	Liquid/solid
Odor	Garlic-like odor
Chemical formula	$C_6H_{12}Cl_2S_2$
Molecular weight	219.20
Chemical structure	$ClCH_2CH_2-S-CH_2CH_2-S-CH_2CH_2Cl$
Melting point	56°C
Boiling point	328.7°C
Flash point	144.9°C
Decomposition temperature	No data available
Vapor density	7.6 (air = 1)
Liquid density	1.27 at 25°C
Solid density	No data available

Vapor pressure	0.000357 mmHg at 25˚C
Volatility	0.412 mmHg
Solubility	Slight in H_2O

Reactivity

Hydrolysis products	Cyclic sulfonium ions were observed in hydrolysis
Rate of hydrolysis	No data available
Stability	No data available
Storage stability	No data available
Decomposition	No data available
Polymerization	No data available

Toxicity

LD_{50} *(skin)*	No data available
LCt_{50} *(respiratory)*	11 mg/m^3 (rat)
LCt_{50} *(percutaneous)*	No data available
ICt_{50} *(respiratory)*	No data available
ICt_{50} *(percutaneous)*	No data available
Rate of detoxification	No data available
Skin and eye toxicity	No data available
Rate of action	No data available
Overexposure effects	No data available

Safety

Protective gloves	Protective mask and permeable protective clothing for vapor and small droplets; impermeable protective clothing for protection against large droplets, splashes, and smears.
Eye protection	HTH, STB, household bleach, fire or DS2. Decontaminate liquid agent on the skin with the M258A1, M258, or M291 skin decontaminating kit. Decontaminate individual equipment with the M280 individual equipment decontamination kit.
Other	Heavily splashed Q liquid persists 1–2 days under average weather conditions, and a week or more under very cold conditions. It persists in water due to poor solubility.
Emergency procedures	Delayed-action casualty agent.

Military Significant Information

Field protection	Protective mask and permeable protective clothing for Q vapor and small droplets; impermeable protective clothing for protection against large droplets, splashes, and smears.
Decontamination	STB, fire, or DS2. Decontaminate liquid agent on the skin with the M258A1, M258, or M291 skin decontaminating kit. Decontaminate individual equipment with the M280 individual equipment decontamination kit.
Persistency	Depends upon the amount of contamination by liquid, the munition used, the nature of the terrain and the soil, and the weather conditions. Heavily splashed liquid persists 1–2 days or more in concentrations that produce casualties of military significance under average weather conditions, and a week to months under very cold conditions. Q on soil remains vesicant for about 2 weeks. Persistency in running water is only a few days, while persistency in stagnant water can be several months.
Use	Delayed-action casualty agent. Not authorized for military use.

1.15. Sulfur Mustard Mixture – HT

HT is a mixture of 60% HD and 40% agent T. It is expected that the effects of HT would encompass of both HD and T. Both HD and T are alkylating agents. T is a sulfur and chlorine compound similar in structure to HD. HT has a strong blistering effect, has a longer duration of effectiveness, is more stable, and has a lower freezing point than HD. Its low volatility makes effective vapor concentrations in the field difficult to obtain. Properties are essentially the same as those of HD.

Informational

Designation	HT
Class	Blister agent
Type	B – persistent
Chemical name	HD: bis-(2-chloroethyl) sulfide T: bis-[2(2-chloroethylthio)ethyl] ether
CAS number	Mixture: [172672-28-5]; HD: [505-60-2]; T: [63918-89-8]

Chemical and Physical Properties

Appearance	Clear yellowish liquid
Odor	Slight garlic- or mustard-like odor

Chemical formula	HD: $C_4H_8Cl_2S$; T: $C_8H_{16}Cl_2OS_2$
Molecular weight	HD: 159.08; T: 263.3; 189.4 based on 60:40 mixture
Chemical structure	$Cl-CH_2CH_2-S-CH_2CH_2-Cl$ $+$ $ClCH_2CH_2SCH_2CH_2-O-CH_2CH_2SCH_2CH_2Cl$
Melting point	0.0–1.3°C for 60:40 mixture
Boiling point	>228°C
Flash point	100°C
Decomposition temperature	165–185°C
Vapor density	6.92 (air = 1) for 60:40 mixture
Liquid density	1.27 g/cm^3 at 25°C
Solid density	No data available
Vapor pressure	0.104 mmHg at 25°C
Volatility	831 mg/m^3 at 25°C
Solubility	Practically insoluble in water. Soluble in most organic solvents.

Reactivity

Hydrolysis products	Hydrochloric acid and thiodiglycol
Rate of hydrolysis	See HD
Stability	Stable at ambient temperatures; decomposition temperature is 165–185°C.
Storage stability	Pressure develops in steel. Rapidly corrosive to brass at 65°C; will corrode steel at 0.001 in. of steel per month at 65°C.
Decomposition	HT will hydrolyze to form hydrochloric acid, thiodiglycol, and bis-(2-(2-hydroxyethylthio) ethyl ether.
Polymerization	Will not occur

Toxicity

LD_{50} *(skin)*	No data available
LCt_{50} *(respiratory)*	No data available
LCt_{50} *(percutaneous)*	No data available
ICt_{50} *(respiratory)*	No data available
ICt_{50} *(percutaneous)*	No data available

Rate of detoxification	Very low, see HD for more detail.
Skin and eye toxicity	Eyes are very susceptible to low concentrations. Incapacitating effects by skin absorption require higher concentrations than does eye injury. HT applied to the skin appears to be more active than HD.
Rate of action	No data available
Overexposure effects	HD is a vesicant (blister agent) and alkylating agent producing cytotoxic action on the hematopoietic (blood-forming) tissues which are especially sensitive. The rate of detoxification of HD in the body is very slow, and repeated exposures produce a cumulative effect. It causes blisters, irritates the eyes, and it is toxic when inhaled. HD has been determined to be a human carcinogen by the International Agency for Research on Cancer.

Safety

Protective gloves	Mandatory; wear butyl toxicological agent protective gloves (M3, M4, gloveset).
Eye protection	Wear chemical goggles as a minimum; use goggles and face shield for splash hazard.
Other	Wear gloves and lab coat with M9 or M17 mask readily available for general lab work. In addition, wear daily clean smock, foot covers, and head cover when handling contaminated lab animals.
Emergency procedures	**Inhalation:** Remove victim from the source immediately; administer artificial respiration if breathing has stopped; administer oxygen if breathing is difficult; seek medical attention immediately.
	Eye Contact: Speed in decontaminating the eyes is absolutely essential; remove victim from the liquid source, flush the eyes immediately with water by tilting the head to the side, pulling the eyelids apart with the fingers, and pouring water slowly into the eyes; do not cover eyes with bandages; but if necessary, protect eyes by means of dark or opaque goggles; seek medical attention immediately.
	Skin Contact: Don respiratory protective masks and gloves; remove victim from agent source immediately; flush skin and clothes with 5% solution of sodium hypochlorite or liquid household bleach within 1 min; cut and remove contaminated clothing; flush contaminated skin area again with 5% sodium hypochlorite solution; then wash contaminated skin area with soap and water; seek medical attention immediately.
	Ingestion: Do not induce vomiting; give victim milk to drink; seek medical attention immediately.

Military Significant Information

Field protection	Protective mask and permeable protective clothing for vapor and small droplets; impermeable protective clothing for protection against large droplets, splashes, and smears.
Decontamination	HTH, STB, household bleach, fire or DS2. Decontaminate liquid agent on the skin with the M258A1, M258, or M291 skin decontaminating kit. Decontaminate individual equipment with the M280 individual equipment decontamination kit.
Persistency	The persistency of HT is somewhat longer than the duration of effectiveness of HD. Heavily splashed HD liquid persists 1–2 days under average weather conditions, and a week or more under very cold conditions. It persists in water due to poor solubility.
Use	Delayed-action casualty agent.

Chapter 2

Blood Agents

The body absorbs these chemical agents, including the cyanide group, primarily by breathing. They poison an enzyme called cytochrome oxidase, blocking the use of oxygen in every cell in the body. Thus, these agents prevent the normal transfer of oxygen from the blood to body tissues. The lack of oxygen rapidly affects all body tissues, especially that of the central nervous system (CNS). Because of the rapidity of the action of cyanide, treatment must be immediate. The general principles of therapy are to remove cyanide from the enzyme cytochrome oxidase and then to detoxify cyanide and remove it from the body. Because nerve tissue is particularly sensitive to the effects of cyanide, there is the possibility of a lingering partial paralysis that may be seen following treatment during the acute phase, and could still be present several weeks post exposure.

Most blood agents are cyanide-containing compounds, absorbed into the body primarily by breathing. AC and CK are the important agents in this group. Blood agents are highly volatile and, therefore, nonpersistent even at very low temperatures. These agents can be dispersed by artillery shell, mortar, rocket, aircraft spray, or bomb. AC has an odor like bitter almonds; CK is somewhat more pungent. The odor of CK often goes unnoticed because CK is so irritating to the eyes, nose, and respiratory tract. At high concentrations both compounds cause effects within seconds and death within minutes in unprotected personnel. Cyanogen chloride also acts as a choking agent. The standard protective mask gives adequate protection against field concentrations.

2.1. Arsine – SA

SA is a gas with a mild, garlic-like odor. It is used as a delayed-action casualty agent that interferes with the functioning of the blood and damages the liver and kidneys. Slight exposure causes headache and uneasiness. Increased exposure causes chills, nausea, and vomiting. Severe exposure damages blood, causing anemia. It is a carcinogen. The protective mask provides adequate protection.

Informational

Designation	SA
Class	Blood agent
Type	A – nonpersistent
Chemical name	Arsenic trihydride
CAS number	[7784-42-1]

Chemical and Physical Properties

Appearance	Colorless gas
Odor	Mild, garlic-like
Chemical formula	AsH_3
Molecular weight	77.93

Chemical structure

$$\begin{array}{c} H \\ | \\ As \\ \diagup \quad \diagdown \\ H \qquad H \end{array}$$

Melting point	−116°C
Boiling point	−62.5°C
Flash point	No data available
Decomposition temperature	280°C
Vapor density	2.69 (air = 1)
Liquid density	1.34 g/cm^3 at 20°C
Solid density	No data available
Vapor pressure	11,100 mmHg at 20°C. This high vapor pressure means that SA is difficult to liquefy and to store.
Volatility	30,900,000 mg-min/m^3 at 20°C. This by far the highest volatility found among the compounds considered for tactical use as chemical agents. This fact, coupled with a relatively low latent heat of vaporization, qualifies SA as the most rapidly dispersing chemical agent.
Solubility	No data available

Reactivity

Hydrolysis products	Arsenic acids and a hydride containing fewer hydrogen atoms than SA itself.

Rate of hydrolysis	Rapid, but reaches an equilibrium condition quickly. (Under certain conditions SA forms a solid product with water that decomposes at 30°C.)
Stability	Stable
Storage stability	Not stable in uncoated containers. Metals catalyze decomposition of SA. Reacts slowly with copper, brass, and nickel. Contact with other metals may also decompose it.
Decomposition	Hydrogen, arsenic and arsenic trioxide at about 232°C.
Polymerization	Does not occur.

Toxicity

LD_{50} (skin)	No data available
LCt_{50} (respiratory)	5,000 mg-min/m^3 (it is estimated that 2 mg of SA per kilogram of body weight would be lethal to humans).
LCt_{50} (percutaneous)	No data available
ICt_{50} (respiratory)	2,500 mg-min/m^3
ICt_{50} (percutaneous)	No data available
Rate of detoxification	Not rapid enough to be of importance
Skin and eye toxicity	None
Rate of action	Effects are delayed from 2 h to as much as 11 days.
Overexposure effects	The symptoms of inhalation of this mixture are not well known. However, *Arsine is an extremely toxic gas* that destroys red blood cells and can cause widespread organ injury. Inhalation may cause headache, dilirium, nausea, vomiting, general malaise, tightness in the chest, and pain in the abdomen and loins. Arsine may discolor urine to red or a darkened color, and the skin to a bronze or jaundiced color. Symptoms may not occur until several hours after exposure.

Safety

Protective gloves	Neoprene, butyl rubber, PVC, polyethylene, or Teflon
Eye protection	Gas-tight safety goggles or full-face respirator
Other	Use positive pressure air line with mask or self-contained breathing apparatus for exposure limits and emergency use. Air purifying respirators may not afford an adequate level of protection.
Emergency procedures	**Inhalation:** Prompt medical attention is mandatory in all cases of exposure. Rescue personnel should be equipped with self-contained breathing apparatus and be aware of extreme fire and explosion hazard. Regard anyone exposed

as having a potentially toxic dose. Quick removal from the contaminated area is most important. Conscious persons should be assisted to an uncontaminated area and inhale fresh air. Unconscious persons should be moved to an uncontaminated area, and given artificial resuscitation and supplemental oxygen if they are not breathing. Further treatment should be symptomatic and supportive. Advise physician of the toxic properties of arsine as a powerful hemolytic agent.

Eye Contact: Flush contaminated eye(s) with copious quantities of water. Part eyelids to assure complete flushing. Continue for a minimum of 30 min. See a physician for follow-up treatment as soon as possible.

Skin Contact: Flush affected area with copious quantities of water. Remove affected clothing as rapidly as possible. Give water by mouth to dilute. Never give anything by mouth to an unconscious person. Call your local poison control center for advice, giving identity of chemical.

Military Significant Information

Field protection	Protective mask
Decontamination	None required
Persistency	Short
Use	Delayed-action casualty agent

2.2. Cyanogen Chloride – CK

Cyanogen chloride irritates the eyes and respiratory tract, even in low concentrations. Acute exposure produces intense irritation of the lungs characterized by coughing and breathing problems, which may quickly lead to a pulmonary edema. Inside the body, cyanogen chloride converts to hydrogen cyanide, which inactivates the enzyme cytochrome oxidase, preventing the utilization of oxygen by the cells. The general action of CK, interference with the use of oxygen in the body, is similar to that of AC. However, CK differs from AC in that it has strong irritating and choking effects and slows breathing.

Informational

Designation	CK
Class	Blood agent
Type	A – nonpersistent
Chemical name	Cyanogen chloride
CAS number	[506-77-4]

Chemical and Physical Properties

Appearance	Cyanogen chloride is a colorless gas. CK is a liquid at temperatures below 55°F.
Odor	Sharp, pepperish odor similar to that of most tear gasses. The odor of CK often goes unnoticed because it is so irritating to the mucous membranes.
Chemical formula	CClN
Molecular weight	61.48
Chemical structure	$N \equiv C - Cl$
Melting point	−6.9°C
Boiling point	12.8°C
Flash point	Does not flash
Decomposition temperature	>100°C
Vapor density	2.1 (air = 1)
Liquid density	1.18 g/cm^3 at 20°C
Solid density	No data available
Vapor pressure	1,000 mmHg at 25°C
Volatility	2,600,000 mg/m^3 at 12.8°C
Solubility	Slightly soluble in water; dissolves readily in alcohol, carbon disulfide, acetone, benzene, carbon tetrachloride, chloropicrin, HD, and AC.

Reactivity

Hydrolysis products	Hydrochloric acid and cyanic acid.
Rate of hydrolysis	Very low
Stability	Unstable
Storage stability	Unstable; polymerizes without stabilizer; stable for less than 30 days in canister munitions; will polymerize to form the solid cyanuric chloride which is corrosive and may explode.
Decomposition	2,4,6-Trichloro-*s*-Triazine which can polymerize violently.
Polymerization	Hazardous polymerization may occur; avoid high temperature storage and moisture.

Toxicity

LD$_{50}$ (skin)	No data available
LCt$_{50}$ (respiratory)	11,000 mg-min/m^3

LCt_{50} *(percutaneous)*	No data available
ICt_{50} *(respiratory)*	7,000 mg-min/m^3
ICt_{50} *(percutaneous)*	No data available
Rate of Detoxification	0.02–0.1 mg/kg/min
Skin and eye toxicity	Too low to be of military importance; highly irritating to eyes, upper respiratory tract, and lungs. CK can cause dryland drowning.
Rate of action	Immediate intense irritation. The systemic effect of CK is believed to arise from its conversion to AC in the body. In general, CK may be considered a rapid-acting chemical agent.
Overexposure effects	CK is absorbed through the skin and mucosal surfaces and is dangerous when inhaled because toxic amounts are absorbed through bronchial mucosa and alveoli. It is similar in toxicity and mode of action to AC, but is much more irritating. CK can cause a marked irritation of the respiratory tract, hemorrhagic exudate of the bronchi and trachea as well as pulmonary edema. It is improbable that anyone would voluntarily remain in areas with a high enough concentration to exert a typical nitrile effect. The liquid form will burn skin and eyes. Long-term exposure will cause dermatitis, loss of appetite, headache, and upper respiratory irritation in humans.

Safety

Protective gloves	Wear butyl or neoprene rubber gloves.
Eye protection	Wear chemical safety goggles if dust or solutions of cyanide salts may come into contact with the eye; wear full-length face shields with forehead protection if dusts, molten salts, or solutions of cyanide salts contact the face.
Other	Wear appropriate chemical cartridge respirator depending on the amount of exposure; rescue personnel should be equipped with self-contained breathing apparatus; have available and use as appropriate rubber suits, full-body chemical suits, safety shoes, safety shower, and eyewash fountain.
Emergency procedures	**Inhalation:** If the patient is conscious, direct first aid and medical treatment toward the relief of any pulmonary symptoms; put patient immediately at bed rest with head slightly elevated; seek medical attention immediately; administer oxygen if there is any dyspnea or evidence of pulmonary edema; in case of long exposures, combined therapy, with oxygen plus amyl nitrite inhalations and artificial respiration is recommended.
	Eye Contact: Flush affected areas with copious amounts of water immediately; hold eyes open while flushing.

Skin Contact: Wash skin promptly to remove the cyanogen chloride; remove all contaminated clothing, including shoes; do not delay.

Ingestion: Give victim water or milk; do not induce vomiting.

Military Significant Information

Field protection	Protective mask. CK will break or penetrate a protective mask canister or filter element more readily than most other agents. A very high concentration may overpower the filter; high dosages will break down its protective ability.
Decontamination	None required under field conditions.
Persistency	Short; vapor may persist in jungle and forest for some time under suitable weather conditions.
Use	Quick-acting casualty agent. Used for degradation of canisters or filter elements in protective mask.

2.3. Hydrogen Cyanide – AC

Hydrogen cyanide is a fast acting, highly poisonous material. It may be fatal if inhaled, swallowed, or absorbed through the skin. It is an extremely hazardous liquid and vapor under pressure. With prompt treatment following overexposure, recovery is normally quick and complete. AC inactivates the enzyme cytochrome oxidase, preventing the utilization of oxygen by the cells. The toxic hazard is high for inhalation, ingestion, and skin and eye exposure, but AC is primarily an inhalation hazard due to its high volatility.

Informational

Designation	AC
Class	Blood agent
Type	A – nonpersistent
Chemical name	Hydrogen cyanide
CAS number	[74-90-8]

Chemical and Physical Properties

Appearance	Pure AC is a nonpersistent, colorless liquid that is highly volatile.
Odor	It has a faint odor similar to bitter almonds or peachy kernels that sometimes cannot be detected even at lethal concentrations.

Chemical formula	HCN
Molecular weight	27.03
Chemical structure	$N \equiv C - H$
Melting point	$-13.3°C$
Boiling point	$25.7°C$
Flash point	$-18°C$
Decomposition temperature	$>65.5°C$. Forms explosive polymers on standing. Stabilized material can be stored up to 65°C.
Vapor density	1.007 at 25.7°C (air = 1.0) 0.990 at 20°C 0.978 at 0°C 0.93 at $-17.8°C$
Liquid density	0.687 g/cm^3 at 20°C 0.716 g/cm^3 at 0°C
Solid density	No data available
Vapor pressure	742 mmHg at 25°C 612 mmHg at 20°C 265 mmHg at 0°C
Volatility	1,080,000 mg/m^3 at 25°C 441,000 mg/m^3 at 0°C 37,000 mg/m^3 at $-40°C$
Solubility	Highly soluble and stable in water and alcohol; soluble in ether, glycerine, chloroform, and benzene.

Reactivity

Hydrolysis products	Ammonia and formic acid.
Rate of hydrolysis	Low under field conditions.
Stability	Unstable with heat, alkaline materials, and water. Do not store wet AC; may react violently with strong mineral acids; experience shows mixtures with about 20% or more sulfuric acid will explode; effects with other acids are not quantified, but strong acids like hydrochloric or nitric would probably react similarly.
Storage stability	Unstable except when very pure. Forms explosive polymer on long standing. Will stabilize with addition of small amount of phosphoric acid or sulfur dioxide. Stabilized material can be stored up to 65°C.
Decomposition	See polymerization
Polymerization	Can occur violently in the presence of heat, alkaline materials, or moisture. Once initiated, polymerization becomes uncontrollable since the reaction is autocatalytic,

producing heat and alkalinity; confined polymerization can cause a violent explosion. AC is stabilized with small amounts of acid to prevent polymerization; it should not be stored for extended periods unless routine.

Toxicity

LD_{50} (skin)	100 mg/kg (liquid)
LCt_{50} (inhale, 0.5 min)	2,000 mg-min/m^3
LCt_{50} (inhale, 30 min)	20,600 mg-min/m^3
ICt_{50}	Varies with concentration
Rate of detoxification	Rapid; 0.017 mg/kg/min
Skin and eye toxicity	Moderate
Rate of action	Very rapid. Incapacitation occurs within 1 or 2 min of exposure to an incapacitating or lethal dose. Death can occur within 15 min after receiving a lethal dose.
Overexposure effects	AC poisoning causes a deceptively healthy pink to red skin color. However, if physical injury or lack of oxygen is involved, the skin color may be bluish. Human health effects of overexposure by inhalation, ingestion, or skin contact may include nonspecific symptoms such as reddening of the eyes, flushing of the skin, nausea, headaches, dizziness, rapid respiration, vomiting, drowsiness, drop in blood pressure, rapid pulse, weakness, and loss of consciousness; CNS stimulation followed by the CNS depression, hypoxic convulsions, and death due to respiratory arrest; temporary alteration of the heart's electrical activity with irregular pulse, palpitations, and inadequate circulation. Higher AC inhalation exposures may lead to fatality. In a few cases, disturbances of vision or damage to the optic nerve or retina have been reported, but the exposures have been acute and at lethal or near-lethal concentrations. Skin permeation can occur in amounts capable of producing systemic toxicity. There are no reports of human sensitization.

Safety

Protective gloves	Wear butyl or neoprene rubber gloves.
Eye protection	Wear chemical splash goggles as a minimum.
Other	Have available and use as appropriate – rubber suits and gloves; full-body chemical suit; self-contained breathing air

supply; HCN detector; first aid and medical treatment
supplies, including oxygen resuscitators.

Emergency procedures

Inhalation: Remove patient to fresh air, and lay patient
down; administer oxygen and amyl nitrite; keep patient
quiet and warm; even with inhalation poisoning, thoroughly check clothing and skin to assure no cyanide is
present; seek medical attention immediately.

Eye Contact: Flush eyes immediately with plenty of water;
remove contaminated clothing; keep patient quiet and
warm; seek medical attention immediately.

Skin Contact: Wash skin promptly to remove the cyanide
while removing all contaminated clothing, including shoes;
do not delay; skin absorption can occur from cyanide dust,
solutions, or HCN vapor; absorption is slower than inhalation, usually measured in minutes compared to seconds;
HCN is absorbed much faster than metal cyanides from
solutions such as sodium, potassium or copper cyanide solutions; even after washing the skin, watch the patient for at
least 1–2 h because absorbed cyanide can continue to work
into the bloodstream; wash clothing before reuse and
destroy contaminated shoes.

Ingestion: Give patient one pint of 1% sodium thiosulfate
solution (or plain water) immediately by mouth and induce
vomiting; repeat until vomit fluid is clear; never give anything by mouth to an unconscious person; give oxygen;
seek medical attention immediately.

Military Significant Information

Field protection

Protective mask. Liquid AC can penetrate skin but, because
AC has a high LCt_{50} and because liquid AC is not likely to
be encountered in the field, protective clothing is required
only in unusual situations.

Decontamination

None required under field conditions.

Persistency

Short; the agent is highly volatile, and in the gaseous state it
dissipates quickly in the air.

Use

Quick-acting casualty agent suitable for surprise attacks.

Chapter 3

Choking Agents

Choking agents injure an unprotected person chiefly in the respiratory tract (the nose, the throat, and particularly the lungs). In extreme cases membranes swell, lungs become filled with liquid, and death results from lack of oxygen; thus, these agents "choke" an unprotected person. Fatalities of this type are called "dryland drownings."

The toxic action of phosgene is typical of choking agents. Phosgene is the most dangerous member of this group and the only one considered likely to be used in the future. Phosgene was used for the first time in 1915, and it accounted for 80% of all chemical fatalities during World War I.

During and immediately after exposure, there is likely to be coughing, choking, a feeling of tightness in the chest, nausea, and occasionally vomiting, headache, and lachrymation. The presence or absence of these symptoms is of little value in immediate prognosis. Some patients with severe coughs fail to develop serious lung injury, while others with little sign of early respiratory tract irritation develop fatal pulmonary edema. A period follows during which abnormal chest signs are absent and the patient may be symptom-free. This interval commonly lasts 2–24 h but may be shorter. It is terminated by the signs and symptoms of pulmonary edema. These begin with cough (occasionally substernally painful), dyspnea, rapid shallow breathing, and cyanosis. Nausea and vomiting may appear. As the edema progresses, discomfort, apprehension, and dyspnea increase and frothy sputum develops. The patient may develop shock-like symptoms, with pale, clammy skin, low blood pressure, and feeble, rapid heartbeat. During the acute phase, casualties may have minimal signs and symptoms and the prognosis should be guarded. Casualties may very rapidly develop severe pulmonary edema. If casualties survive more than 48 h the recovery is good.

3.1. Chlorine – Cl

The first large-scale use of chemical agents came in World War I on 22 April 1915, when the Germans released chlorine gas against the Allied positions at Ypres, Belgium. The gas was very effective, killing 5,000 and scaring 10,000, but also prevented continuance of the assault. There were other gas attacks by both combatant forces during World War I, and it is well documented that approximately one third of all American casualties in this conflict were due to chemical agent attacks.

Informational

Designation	CL
Class	Choking agent
Type	A – nonpersistent
Chemical name	Chlorine
CAS number	[7782-50-5]

Chemical and Physical Properties

Appearance	Greenish yellow gas or amber liquid
Odor	Pungent, suffocating bleach-like odor
Chemical formula	Cl_2
Molecular weight	70.9
Chemical structure	Cl—Cl
Melting point	$-101°C$
Boiling point	$-34°C$
Flash point	$>600°C$
Decomposition temperature	No data available
Vapor density	2.5 (air = 1)
Liquid density	1.393 g/cm^3 at 25°C 1.468 g/cm^3 at 0°C
Solid density	No data available
Vapor pressure	5168 mmHg at 21°C
Volatility	2.19×10^7 mg/m^3 at 25°C
Solubility	Slightly soluble in water; 0.63 g/100 g H_2O at 25°C

Reactivity

Hydrolysis products	Forms hypochlorous acid and hydrochloric acid.
Rate of hydrolysis	Slow

Stability	Chlorine is extremely reactive. Liquid or gaseous chlorine can react violently with many combustible materials and other chemicals including water. Metal halides, carbon, finely divided metals, and sulfides can accelerate the rate of chlorine reactions. Hydrocarbon gases, e.g., methane, acetylene, ethylene, or ethane, can react explosively if initiated by sunlight or a catalyst. Liquid or solid hydrocarbons, e.g., natural or synthetic rubbers, naphtha, turpentine, gasoline, fuel gas, lubricating oils, greases, or waxes, can react violently. Metals, e.g., finely powdered aluminum, brass, copper, manganese, tin, steel, and iron, can react vigorously or explosively with chlorine. Nitrogen compounds, e.g., ammonia and other nitrogen compounds, can react with chlorine to form highly explosive nitrogen trichloride. Nonmetals, e.g., phosphorous, boron, activated carbon, and silicon can ignite on contact with gaseous chlorine at room temperature. Certain concentrations of chlorine–hydrogen can explode by spark ignition. Chlorine is strongly corrosive to most metals in the presence of moisture. Copper may burn spontaneously. Chlorine reacts with most metals at high temperatures. Titanium will burn at ambient temperature in the presence of dry chlorine.
Storage stability	Stable at normal temperatures and pressure.
Decomposition	Hydrogen chloride may form from chlorine in the presence of water vapor.
Polymerization	Will not occur

Toxicity

LD_{50} *(skin)*	<50 mg/kg
LCt_{50} *(respiratory)*	No data available
LCt_{50} *(percutaneous)*	No data available
ICt_{50} *(respiratory)*	No data available
ICt_{50} *(percutaneous)*	No data available
Rate of detoxification	No data available
Skin and eye toxicity	No data available
Rate of action	Rapid
Overexposure effects	*Acute:* Low concentrations can cause itching and burning of the eyes, nose, throat, and respiratory tract. At high concentrations chlorine is a respiratory poison. Irritant effects become severe and may be accompanied by tearing of the eyes, headache, coughing, choking, chest pain, shortness of breath, dizziness, nausea, vomiting, unconsciousness, and death. Bronchitis and accumulation of fluid in the lungs (chemical pneumonia) may occur hours after exposure to

high levels. Liquid as well as vapor contact can cause irritation, burns, and blisters. Ingestion can cause nausea and severe burns of the mouth, esophagus, and stomach.

Chronic: Prolonged or repeated overexposure may result in many or all of the effects reported for acute exposure (including pulmonary function effects).

Safety

Protective gloves	Wear appropriate protective gloves to prevent any possibility of contact with skin; butyl and neoprene rubber gloves are preferred.
Eye protection	Wear splash-proof or dust-resistant safety goggles and a faceshield to prevent contact with substance.
Other	Wear respirators based on contamination levels found, must not exceed the working limits of the respirator and must be jointly approved by NIOSH; employer should provide an eye wash fountain and quick drench shower for emergency use.
Emergency procedures	**Inhalation:** If adverse effects occur, remove to uncontaminated area. Give artificial respiration if not breathing. If breathing is difficult, oxygen should be administered. Get immediate medical attention.
	Eye Contact: Immediately flush eyes with plenty of water for at least 15 min. Then get immediate medical attention.
	Skin Contact: Wash skin with soap and water for at least 15 min while removing contaminated clothing and shoes. Get immediate medical attention. Thoroughly clean and dry contaminated clothing and shoes before reuse. Destroy contaminated shoes.
	Ingestion: Contact local poison control center or physician immediately. Never make an unconscious person vomit or drink fluids. Give large amounts of water or milk. Allow vomiting to occur. When vomiting occurs, keep head lower than hips to help prevent aspiration. If person is unconscious, turn head to side. Get medical attention immediately.

Military Significant Information

Field protection	Protective mask
Decontamination	For confined areas, aeration is essential; not required in the field.
Persistency	Short. Vapor may persist for some time in low places under calm or light winds and stable atmospheric conditions.
Use	Not authorized for US military use.

3.2. Diphosgene – DP

Diphosgene (DP) has a much higher boiling point than CG. Because DP has a stronger tearing effect, it has less surprise value than CG when used on troops. Its lower volatility (vapor pressure) adds to the difficulty of setting up an effective surprise concentration. DP can produce delayed or immediate casualties, depending upon the dosage received. Because the body converts DP to CG, the physical effects are the same for both agents. Immediate symptoms may follow exposure to a high concentration of DP; a delay of 3 h or more may elapse before exposure to a low concentration causes any ill effects.

Informational

Designation	DP
Class	Choking agent
Type	A – nonpersistent
Chemical name	Trichloromethyl chloroformate
CAS number	[503-38-8]

Chemical and Physical Properties

Appearance	Colorless oily liquid
Odor	New mown hay, grain, or green corn
Chemical formula	$C_2Cl_4O_2$
Molecular weight	197.85
Chemical structure	$Cl-C(=O)-O-CCl_3$
Melting point	−57°C
Boiling point	127–128°C
Flash point	None
Decomposition temperature	300–350°C (yields two molecules of CG, which decomposes at 800°C)
Vapor density	6.8 (air = 1)
Liquid density	1.65 g/cm^3 at 20°C
Solid density	No data available
Vapor pressure	4.2 mmHg at 20°C 1 mmHg at 0°C

Volatility	12,000 mg/m^3 at 0°C
	45,000 mg/m^3 at 20°C
	270,000 mg/m^3 at 51.7°C
Solubility	Limited in water; good in organic solvents.

Reactivity

Hydrolysis products	Hydrochloric acid and carbon dioxide
Rate of hydrolysis	Slow at ordinary temperatures
Stability	Metals act as catalyzers in conversion to CG. Combines vigorously or explosively with water.
Storage stability	Unstable because of conversion to CG.
Decomposition	Hydrogen chloride, phosgene, irritating and toxic fumes and gases, hydrogen gas, and hydrogen gas.
Polymerization	No data available

Toxicity

LD$_{50}$ (skin)	No data available
LCt$_{50}$ (respiratory)	3,000–3,200 mg-min/m^3
LCt$_{50}$ (percutaneous)	No data available
ICt$_{50}$ (respiratory)	1,600 mg-min/m^3
ICt$_{50}$ (percutaneous)	No data available
Rate of detoxification	Not detoxified, cumulative
Skin and eye toxicity	No effect on skin, slight tearing effect
Rate of action	Delayed. Although immediate symptoms may follow exposure to a high concentration of DP, a delay of 3 h or more may elapse before exposure to a low concentration causes any ill effect.
Overexposure effects	No data available

Safety

Protective gloves	Wear appropriate protective gloves to prevent skin exposure.
Eye protection	Wear appropriate protective eyeglasses or chemical safety goggles.
Other	Wear respirators based on contamination levels found, must not exceed the working limits of the respirator and must be

jointly approved by NIOSH; employer should provide an eye wash fountain and quick drench shower for emergency use.

Emergency procedures

Inhalation: Get medical aid immediately. Remove from exposure and move to fresh air immediately. If not breathing, give artificial respiration. If breathing is difficult, give oxygen. Do *not* use mouth-to-mouth resuscitation.

Eye Contact: Get medical aid immediately. Do *not* allow victim to rub eyes or keep eyes closed. Extensive irrigation with water is required (at least for 30 min).

Skin Contact: Get medical aid immediately. Immediately flush skin with plenty of water for at least 15 min while removing contaminated clothing and shoes. Wash clothing before reuse. Destroy contaminated shoes.

Ingestion: Do not induce vomiting. If victim is conscious and alert, give 2–4 cupfuls of milk or water. Never give anything by mouth to an unconscious person. Get medical aid immediately.

Military Significant Information

Field protection   Protective mask

Decontamination   Live stream, ammonia, and aeration for confined area; not required in the field.

Persistency   About 30 min to 3 h in summer; 10–12 h in winter.

Use   Delayed- or immediate-action casualty agent, depending upon dosage rate.

3.3. Nitric Oxide – NO

Nitric oxide (NO) is severely irritating to eyes and respiratory system. Effects may be delayed for several hours following exposure. Corrosive. Inhalation may result in chemical pneumonitis and pulmonary edema. Nonflammable. Oxidizer. This product accelerates the combustion of combustible material.

Informational

Designation   NO

Class   Choking agent

Type   A – nonpersistent

Chemical name   Nitric oxide

CAS number   [10102-43-9]

Chemical and Physical Properties

Appearance	Colorless gas, reddish brown in air
Odor	Suffocating odor
Chemical formula	NO
Molecular weight	30.0
Chemical structure	$N=O$
Melting point	$-163.6°C$
Boiling point	$-151.9°C$
Flash point	Does not flash
Decomposition temperature	No data available
Vapor density	1.04 (air = 1)
Liquid density	No data available
Solid density	No data available
Vapor pressure	25,992 mmHg at 25°C
Volatility	No data available
Solubility	Slight in water. Reacts to form nitric acid

Reactivity

Hydrolysis products	Forms nitric and nitrous acids.
Rate of hydrolysis	Moderate
Stability	Stable. Reacts vigorously with fluorine, fluorine oxides, and chlorine in the presence of moisture. Nitric oxide is non-corrosive and may be used with most common structural materials. However, in the presence of moisture and oxygen, corrosive conditions will develop as a result of the formation of nitric and nitrous acids.
Storage stability	Prior to use, systems to contain nitric oxide must first be purged with an inert gas. Where air contamination cannot be eliminated, stainless steel materials should be used.
Decomposition	Oxidizes in air to form nitrogen dioxide, which is extremely reactive and a strong oxidizer. Upon contact with moisture and oxygen, it produces nitrous and nitric acids.
Polymerization	Will not occur

Toxicity

LD_{50} (skin)	No data available
LCt_{50} (respiratory)	1068 mg/m^3

LCt_{50} *(percutaneous)*	No data available
ICt_{50} *(respiratory)*	No data available
ICt_{50} *(percutaneous)*	No data available
Rate of detoxification	No data available
Skin and eye toxicity	No data available
Rate of action	Rapid
Overexposure effects	Nitric oxide is severely irritating to eyes and respiratory system. Effects may be delayed for several hours following exposure. Nitric oxide vapors are a strong irritant to the pulmonary tract. At high concentrations, initial symptoms of inhalation may be moderate and include irritation to the throat, tightness of the chest, headache, nausea, and gradual loss of strength. Severe symptoms may be delayed (possible for several hours) and include cyanosis, increased difficulty in breathing, irregular respiration, lassitude, and possible eventual death due to pulmonary edema in untreated cases.

Safety

Protective gloves	Protective gloves of rubber or Teflon.
Eye protection	Gas-tight safety goggles or full-face respirator.
Other	Safety shoes and eyewash. Positive pressure air line with full-face mask and escape bottle or self-contained breathing apparatus should be available for emergency use.
Emergency procedures	**Inhalation:** *Prompt medical attention is mandatory in all cases of overexposure. Rescue personnel should be equipped with self-contained breathing apparatus.* Conscious victims should be *carried* (not assisted) to an uncontaminated area and inhale fresh air with supplemental oxygen. Quick removal from the contaminated area is most important. Keep the patient warm, quiet, and under competent medical observation until the danger of delayed pulmonary edema has passed (at least for 72 h). Any physical exertion during this period should be discouraged as it may increase the severity of the pulmonary edema or chemical pneumonitis. Bed rest is indicated. Unconscious persons should be moved to an uncontaminated area, and if breathing has stopped, administer artificial resuscitation and supplemental oxygen. Once respiration has been restored they should be treated as above.
	Eye Contact: Immediately flush with tepid water in large quantities, or with a sterile saline solution. Seek medical attention as soon as possible.
	Skin Contact: Immediately flush with tepid water in large quantities, or with a sterile saline solution. Seek medical attention if blisters or other reactions develop.

Military Significant Information

Field protection	Protective mask
Decontamination	For confined areas, aeration is essential; not required in the field.
Persistency	Short. Vapor may persist for some time in low places under calm or light winds and stable atmospheric conditions.
Use	Not authorized for US military use.

3.4. Perfluoroisobutylene – PFIB

Highly toxic perfluoroisobutylene (PFIB) poses a serious health hazard to the human respiratory tract. PFIB is a thermal decomposition of polytetrafluoroethylene (PTFE), e.g., Teflon. PFIB is approximately 10× as toxic as phosgene. Inhalation of this gas can cause pulmonary edema, which can lead to death. PFIB is included in Schedule 2 of the Chemical Weapons Convention (CWC), the aim of the inclusion of chemicals such as PFIB was to cover those chemicals, which would pose a high risk to the CWC.

Informational

Designation	PFIB
Class	Choking agent
Type	A – nonpersistent
Chemical name	Octafluoroisobutylene
CAS number	[382-21-8]

Chemical and Physical Properties

Appearance	Colorless gas
Odor	Odorless
Chemical formula	C_4F_8
Molecular weight	200.03
Chemical structure	
Melting point	−130°C
Boiling point	7°C

Flash point	−36.4°C
Decomposition temperature	No data available
Vapor density	6.3 (air = 1)
Liquid density	2.1 g/cm^3 at 37°C
Solid density	No data available
Vapor pressure	1740 mmHg at 25°C
Volatility	No data available
Solubility	Decomposes in water

Reactivity

Hydrolysis products	Decomposes to yield fluorophosgene, which in turn decomposes to yield carbon dioxide and hydrogen fluoride.
Rate of hydrolysis	Rapid
Stability	No data available
Storage stability	No data available
Decomposition	Oxidized products containing carbon, fluorine, and oxygen are formed in the air during the thermal decomposition of the fluorocarbon chain.
Polymerization	Does not occur

Toxicity

LD$_{50}$ (skin)	No data available
LCt$_{50}$ (respiratory)	No data available
LCt$_{50}$ (percutaneous)	No data available
ICt$_{50}$ (respiratory)	No data available
ICt$_{50}$ (percutaneous)	No data available
Rate of detoxification	Slow
Skin and eye toxicity	No data available
Rate of action	Rapid
Overexposure effects	**Inhalation:** 2 h LC$_{50}$, rat: 1.05 ppm (extremely toxic). Single and repeated exposure caused cyanosis, impaired lung function, pulmonary edema, and decreased body weight. PFIB is an extremely toxic gas for which inhalation is the most likely route of human exposure. Inhalation exposure may cause severe symptoms of pulmonary edema with wheezing, difficulty in breathing, coughing up sputum, and bluish discoloration of the skin. Coughing and chest pain may

occur initially. However, severe symptoms of pulmonary edema may be delayed for several hours and then become rapidly worse. Overexposure may cause death. Increased susceptibility to the effects of this material may be observed in persons with preexisting disease of the lungs.

Safety

Protective gloves	Protective gloves of rubber or Teflon
Eye protection	Gas-tight safety goggles or full-face respirator.
Other	Safety shoes and eyewash. Positive pressure air line with full-face mask and escape bottle or self-contained breathing apparatus should be available for emergency use.
Emergency procedures	***Inhalation:*** If inhaled, immediately remove to fresh air. If not breathing, give artificial respiration. If breathing is difficult, give oxygen. Call a physician.
	Eye Contact: In case of contact, immediately flush eyes with plenty of water for at least 15 min. Call a physician.
	Skin Contact: Flush area with lukewarm water. Do not use hot water. If frostbite has occurred, call a physician.
	Ingestion: Ingestion is not considered a potential route of exposure.

Military Significant Information

Field protection	Protective mask
Decontamination	For confined areas, aeration is essential; not required in the field.
Persistency	Short. Vapor may persist for some time in low places under calm or light winds and stable atmospheric conditions.
Use	Not authorized for US military use.

3.5. Phosgene – CG

Phosgene was first used in the dye industry in the late 19th century to process colorfast materials. The Germans introduced CG in 1915 for use in World War I. CG, normally a chemical agent with a short duration, was used extensively in World War I. More than 80% of World War I chemical agent fatalities were caused by CG. In the late 1920s, many countries manufactured phosgene as a chemical warfare agent. CG is a severe eye, mucous membrane, and skin irritant. It is highly toxic by inhalation. Two parts per million in air is immediately dangerous to life and health. Being a gas, it is primarily a toxic hazard by inhalation exposure.

Informational

Designation	CG
Class	Choking agent
Type	A – nonpersistent
Chemical name	Carbonic dichloride
CAS number	[75-44-5]

Chemical and Physical Properties

Appearance	CG is foglike in its initial concentration, but it becomes colorless as it spreads.
Odor	It has both a newly mown hay and highly toxic suffocating odor.
Chemical formula	CCl_2O
Molecular weight	98.92
Chemical structure	$$Cl - \overset{\overset{\textstyle O}{\|}}{C} - Cl$$
Melting point	$-128°C$
Boiling point	$7.6°C$
Flash point	None
Decomposition temperature	$800°C$
Vapor density	3.4 (air = 1)
Liquid density	1.373 g/cm^3 at $20°C$ (pure CG) 1.381 g/cm^3 at $20°C$ (plant purity)
Solid density	No data available
Vapor pressure	1,180 mmHg at $20°C$ 555 mmHg at $0°C$ 365 mmHg at $-10°C$
Volatility	528,000 mg/m^3 at $-40°C$ 2,200,000 mg/m^3 at $-10°C$ 4,300,000 mg/m^3 at $7.6°C$
Solubility	Limited in water; decomposes immediately; completely miscible in most organic solvents.

Reactivity

Hydrolysis products	Hydrochloric acid and carbon dioxide

Rate of hydrolysis	Rapid under field conditions. Rain destroys effectiveness. Heavy vegetation, jungle, and forests cause considerable loss by hydrolysis on leafy surfaces.
Stability	Reacts with vapors of sodium with luminescence at about 260°C. Reacts explosively with thionyl chloride or potassium; reacts violently with hexafluoro isopropylidene, amino lithium, ammonia, and strong acids; reacts with tert-butyl azidoformate to form explosive carbide; reacts with 24-hexadiyn-1, 6-diol to form 2, 4-hexadiyn-1, 6-bischloroformate, a shock-sensitive compound; reacts with isopropyl alcohol to form isopropyl chloroformate and hydrogen chloride; thermal decomposition may occur in the presents of iron salts and result in explosion.
Storage stability	Stable in steel containers if CG is dry.
Decomposition	Moist phosgene is very corrosive; it decomposes in the presence of moisture to form hydrochloric acid and carbon monoxide; thermal decomposition may release toxic and/or hazardous gases.
Polymerization	Will not occur

Toxicity

LD_{50} *(skin)*	No data available
LCt_{50} *(respiratory)*	3,200 mg-min/m^3
LCt_{50} *(percutaneous)*	No data available
ICt_{50} *(respiratory)*	1,600 mg-min/m^3
ICt_{50} *(percutaneous)*	No data available
Rate of detoxification	Not detoxified; cumulative
Skin and eye toxicity	Mild eye irritation
Rate of action	Delayed. Immediate irritation in high concentrations. Exposure to low concentration may cause no ill effects for 3 h or more.
Overexposure effects	Phosgene is a corrosive, highly toxic gas used as a delayed-casualty agent resulting in fluid buildup in the lungs ("dry-land drowning"). It affects the upper respiratory tract, skin, and eyes and causes severe respiratory damage as well as burns to the skin and eyes. Acute inhalation may cause respiratory and circulatory failure with symptoms of chills, dizziness, thirst, burning of eyes, cough, viscous sputum, dyspnea, feeling of suffocation, tracheal rhonchi, burning in throat, vomiting, pain in chest, and cyanosis. Rapid progression to pulmonary edema and pneumonia, and death from respiratory and circulatory failure may occur. Pulmonary edema can suddenly occur up to 48 h after exposure. Phosgene is a severe mucous membrane irritant. Chronic inhalation may cause irreversible pulmonary changes

resulting in emphysema and fibrosis. Acute skin contact lesions similar to those of frostbite and burns; it is a severe skin irritant. Chronic skin contact may result in dermatitis. Acute eye contact may result in conjunctivitis, lacrimation, lesions similar to those of frostbite, and burns; chronic eye contact may result in conjunctivitis.

Safety

Protective gloves	Wear appropriate protective gloves to prevent any possibility of contact with skin; butyl and neoprene rubber gloves are preferred.
Eye protection	Wear splash-proof or dust-resistant safety goggles and a faceshield to prevent contact with substance.
Other	Wear respirators based on contamination levels found in the workplace; must not exceed the working limits of the respirator and must be jointly approved by NIOSH; employer should provide an eye wash fountain and quick drench shower for emergency use.
Emergency procedures	*Inhalation:* Remove victim to fresh air; keep individual calm and avoid any unnecessary exertion or movement; maintain airway and blood pressure; trained persons should administer oxygen if breathing is difficult; give artificial respiration if victim is not breathing; seek medical attention immediately.
	Eye Contact: Flush eyes immediately with running water or normal saline for at least 15 min; hold eyelids apart during irrigation; do not delay rinsing to avoid permanent eye injury; seek medical attention immediately.
	Skin Contact: Unlikely that emergency treatment will be required; gently wrap affected part in blankets if warm water is not available or practical to use; allow circulation to return naturally; if adverse effects occur, seek medical attention immediately.
	Ingestion: Treat symptomatically and supportively; if vomiting occurs, keep head lower than hips to prevent aspiration; seek medical attention immediately.

Military Significant Information

Field protection	Protective mask
Decontamination	For confined areas, aeration is essential; not required in the field.
Persistency	Short. Vapor may persist for some time in low places under calm or light winds and stable atmospheric conditions (approximately 30 min in summer; approximately 3 h at $-20°C$).
Use	Delayed-action casualty agent

Chapter 4

Incapacitating Agents

An incapacitant is a chemical agent, which produces a temporary disabling condition that persists for hours to days after exposure to the agent has occurred (unlike that produced by riot control agents). Medical treatment while not essential may in some cases facilitate more rapid recovery. In the narrower sense the term has come to mean those agents that are:

- Highly potent (an extremely low dose is effective) and logistically feasible.
- Able to produce their effects by altering the higher regulatory activity of the central nervous system.
- Of a duration of action lasting hours or days, rather than of a momentary or fleeting action.
- Not seriously dangerous to life except at doses many times the effective dose.
- Not likely to produce permanent injury in concentrations which are militarily effective.

These criteria eliminate many drugs that might otherwise be considered as incapacitants. Opiates and strong sedatives are too dangerous on account of their low margin of safety and milder tranquilizers cause little actual loss of performance capability. Many compounds have been considered as incapacitants and medical staffs must be on the alert to detect and report any unusual clinical appearances. All lethal agents in low doses may produce incapacitating effects and it is possible that new agents for incapacitation may be developed. Agents that produce unconsciousness or induce vomiting may well be developed in the future.

In this chapter, consideration will be given to two categories, which are well known: CNS depressants (anticholinergics) and CNS stimulants (LSD). Although cannabinols and psylocibin, for instance, have been considered in the past, their effective dose is too high for these to be regarded as likely agents for use in the field.

4.1. Agent – BZ

BZ is usually disseminated as an aerosol with the primary route of entry into the body through the respiratory system; the secondary route is through the digestive tract. BZ blocks the action of acetylcholine in both the peripheral and central nervous systems. As such, it lessens the degree and extent of the transmission of impulses from one nerve fiber to another through their connecting synaptic junctions. It stimulates the action of noradrenaline (norepinephrine) in the brain, much as do amphetamines and cocaine. Thus, it may induce vivid hallucinations as it sedates the victim. Toxic delirium is very common.

Informational

Designation	BZ
Class	Psychedelic agents
Type	B – persistent
Chemical name	3-Quinuclidinyl benzilate
CAS number	[6581-06-2]

Chemical and Physical Properties

Appearance	White crystalline solid
Odor	Odorless
Chemical formula	$C_{21}H_{23}NO_3$
Molecular weight	337.4

Chemical structure	

Melting point	167.5°C
Boiling point	320°C
Flash point	246°C
Decomposition temperature	Begins to decompose at about 170°C in air under prolonged heating; is almost completely decomposed after 1–2 h at 200°C. Decomposition rate is both temperature and purity dependent.

Vapor density	11.6 (air = 1)
Liquid density	No data available
Solid density	1.33 g/cm^3
Vapor pressure	0.03 mmHg at 70°C
Volatility	0.5 mg/m^3 at 70°C
Solubility	Slightly soluble in water; soluble in dilute acids, trichloroethylene, warm dimethylformamide, and most organic solvents, such as alcohol and chloroform; insoluble in aqueous alkali. Salts formed with inorganic and organic acids are soluble.

Reactivity

Hydrolysis products	3-Quinuclidinol and benzilic acid
Rate of hydrolysis	Half-life at 25°C is 6.7 h at pH 9.8; 1.8 min at pH 13 and 3–4 weeks in moist air. Half-life at 37°C is 95 h at pH 7.4 and 10 h at pH 9.
Stability	Resistance to air oxidation at ambient temperature (half-life at 25°C at pH 7 and 3–4 weeks). Stable in 0.1 N H$_2$SO$_4$. Thermally stable in solution (half-life more than 2 h at 235°C for pure sample).
Storage stability	Stable in storage and glass containers. BZ lightly attacks aluminum and anodized aluminum after 3 months at 71°C. No effects on steel or stainless steel after 3 months.
Decomposition	Pyrolysis occurs at 170°C after prolonged periods yielding CO, CO$_2$, benzophenone, and benzhydrol; appreciable hydrolysis in acidic or basic solutions occurs yielding 3-quinuclidinol and benzilic acid; BZ is oxidized by hypochlorite at a pH of 1–13.
Polymerization	Will not occur

Toxicity

LD$_{50}$ (skin)	No data available
LCt$_{50}$ (respiratory)	200,000 mg-min/m^3 (estimated)
LCt$_{50}$ (percutaneous)	No data available
ICt$_{50}$ (respiratory)	101 mg-min/m^3
ICt$_{50}$ (percutaneous)	No data available
Rate of detoxification	From an ICt$_{50}$ dose severe effects last 36 h; mild effects last 45 h.
Skin and eye toxicity	No data available

Rate of action   Delayed. Usual onset of symptoms occurs approximately 2 h after aerosol exposure. Depending on inhaled or ingested dosage, symptoms may appear at times ranging from 30 min to 20 h after exposure. Effects from skin contact may appear 36 h later. Dimethylsulfoxide as a "carrier" increases the percutaneous effect by a factor of at least 25.

Overexposure effects   BZ is a very potent psychoactive chemical affecting the CNS as well as the organs of circulation, digestion, salivation, sweating, and vision. Its pharmacological action is similar to that of other anticholinergic drugs (e.g., atropine and scopolamine), but long lasting. Acute exposure produces: increased heart and respiratory rates; mydriasis; mouth, skin, and lip dryness; cycloplegia; high temperature; ataxia; flushing of face and neck; hallucinations; stupor; forgetfulness; and confusion. The initial symptoms after ½–4 h of exposure include: dizziness, mouth dryness, and increased heart rate; secondary symptoms, after 3–5 h of exposure, include: restlessness, involuntary muscular movements, rear vision impairment, and total incapacitation; and final symptoms, after 6–10 h of exposure are psychotropic in nature. After 3–4 days, full recovery from BZ intoxication is expected.

Safety

Protective gloves   Wear butyl (M3 or M4) or neoprene gloves.

Eye protection   Wear protective eyeglass (goggles with hooded ventilation) as a minimum.

Other   Wear maximum protection for nonlab operations consisting of M9 mask and hood, M3 butyl rubber suit, M2A1 butyl boots, M3 or M4 gloves, unimpregnated underwear, or demilitarization protective ensemble (DPE). For specific BZ operations, the local safety office will determine the required level of protective clothing; it will be specified in the local standing operating procedures. Wear lab coats and impervious gloves for lab operations; have masks readily available.

Emergency procedures   **Inhalation:** Remove individual from exposure immediately; start resuscitation and administer oxygen if breathing is irregular or has stopped; seek medical attention immediately.

Eye Contact: Flush eyes with water for at least 15 min; do not rub eyes; seek medical attention immediately.

Skin Contact: Wash from skin and clothing with water; remove any contaminated clothing; seek medical attention immediately.

Ingestion: Do not induce vomiting; seek medical attention immediately.

Military Significant Information

Field protection   The principles applied to the nerve agents apply equally as well to the incapacitating agents. It is possible that such agents will be disseminated by smoke-producing munitions or aerosols, using the respiratory tract as a route of entry. The use of protective mask, therefore, is essential. The skin is usually a much less effective route.

Decontamination   Complete cleansing of the skin with soap and water at the earliest opportunity. If washing is impossible, use the M258A1, M258, or M291 skin decontamination kit. Symptoms may appear as late as 36 h after contact exposure, even if the skin is washed within an hour. In fact, a delay in onset of several hours is typical. Use this time to prepare for the possibility of a widespread outbreak 6–24 h after the attack. Decontaminate bulk quantities of BZ with caustic alcohol solutions. If BZ or other belladonnoids are used as free bases, decontamination will require a solvent, such as 25% ethanol, 0.1 N hydrochloric acid, or 5% acetic acid.

Persistency   Very persistent in soil and water and on surfaces.

Use   Delayed-action incapacitating agent.

Chapter 5

Nerve Agents

Nerve agents are organophosphate ester derivatives of phosphoric acid. They are generally divided into the G-agents, which in the unmodified state are volatile, and the V-agents, which tend to be more persistent. Even G-agents are capable of being thickened with various substances to increase the persistence and penetration of the intact skin. The principal nerve agents are Tabun (GA), Sarin (GB), Soman (GD), and VX.

The G-agents are fluorine- or cyanide-containing organophosphates. In pure form they are colorless liquids. Their solubility in water ranges from complete miscibility for GB to almost total insolubility for GD. They have a weakly fruity odor but in field concentrations are odorless. Clothing gives off G-agents for about 30 min after contact with vapor; consider this fact before unmasking.

The V-agents are sulfur-containing organophosphorous compounds. They are oily liquids with high boiling points, low volatility, and result in high persistency. They are primarily contact hazards. They are exceptionally toxic; the limited amount of vapor they produce is sufficient to be an inhalation hazard. They have very limited solubility in water and are hydrolyzed only minimally. V-agents affect the body in essentially the same manner as G-agents.

The nerve agents are all viscous liquids, not nerve gas per se. However, the vapor pressures of the G-series nerve agents are sufficiently high for the vapors to be lethal rapidly. The volatility is a physical factor of most importance. GB is so volatile that small droplets sprayed from a plane or released from a shell exploding in the air may never reach the ground. This total volatilization means that GB is largely a vapor hazard. At the other extreme agent VX is of such low volatility that it is mainly a liquid contact hazard. Toxicity can occur from the spray falling on one's skin or clothes and from touching surfaces on which the spray has fallen. GD is also mainly a vapor hazard, while GA can be expected to contaminate surfaces for a sufficiently long time to provide a relevant contact hazard.

Thickeners added to GD increase persistence in the field. The thickened agents form large droplets that provide a greater concentration reaching the ground and a greater contact hazard than the unthickened forms.

The relative solubility of these compounds in water and soil is of significance because it relates to their disposition. The ability of GB and GA to mix with water means that water could wash them off surfaces, that these agents can easily contaminate water sources, and that they will not penetrate skin as readily as the more fat-soluble agents VX and GD. G-agents spread rapidly on surfaces, such as skin; VX spreads less rapidly, and the thickened agents very slowly. The moist surfaces in the lungs absorb all the agents very well.

Both the G- and V-agents have the same physiological action on humans. They are potent inhibitors of the enzyme acetylcholinesterase (AChE), which is required for the function of many nerves and muscles in nearly every multicellular animal. Normally, AChE prevents the accumulation of acetylcholine after its release in the nervous system. Acetylcholine plays a vital role in stimulating voluntary muscles and nerve endings of the autonomic nervous system and many structures within the CNS. Thus, nerve agents that are cholinesterase inhibitors permit acetylcholine to accumulate at those sites, mimicking the effects of a massive release of acetylcholine. The major effects will be on skeletal muscles, parasympathetic end organs, and the CNS.

Individuals poisoned by nerve agents may display the following symptoms of difficulty in breathing, drooling and excessive sweating, nausea, vomiting, cramps, and loss of bladder/bowel control, twitching, jerking, and staggering, headache, confusion, drowsiness, coma, and convulsion.

The number and severity of symptoms depend on the quantity and route of entry of the nerve agent into the body. When the agent is inhaled, a prominent symptom is pinpointing of the pupils of the eyes and dimness of vision because of the reduced amount of light entering. However, if exposure has been through the skin or by ingestion of a nerve agent, the pupils may be normal or only slightly to moderately reduced in size. In this event, diagnosis must rely upon the symptoms of nerve agent poisoning other than its effects on the pupils.

Exposure through the eyes produces a very rapid onset of symptoms (usually less than 2–3 min). Respiratory exposure usually results in onset of symptoms in 2–5 min; lethal doses kill in less than 15 min. Liquid in the eye kills nearly as rapidly as respiratory exposure.

Symptoms appear much more slowly from skin absorption. Skin absorption great enough to cause death may occur in 1–2 h. Respiratory lethal dosages kill in 1–10 min, and liquid in the eye kills nearly as rapidly. Very small skin dosages sometimes cause local sweating and tremors but little other effects. Nerve agents are cumulative poisons. Repeated exposure to low concentrations, if not too far apart, will produce symptoms.

Treatment of nerve agent poisoning includes use of the nerve agent antidote (atropine and 2–PAM chloride). Atropine blocks acetylcholine; 2–PAM Cl, reactivates the enzyme AChE. As time passes without treatment the binding of nerve agents to AChE "ages" and the 2–PAM Cl can no longer remove the agent. Certain agents, such as GD, that age rapidly may resist treatment if it is not prompt.

The book would be incomplete if I did not at least mention the subject of novichok (Russian for "newcomer") class of agents.

The earliest information on Russia's novichok chemical weapons program which was codenamed "Foliant," carne just prior to Moscow's signing of the Chemical Weapons Convention (CWC) from two Russian chemists. In the late 1980s and early 1990s, Russia produced several new agents that were reportedly made from chemicals not controlled by the CWC.

Although no details are available, these agents are a novel family of binary nerve agents. Novichok compounds were derived from this new set of unitary agents (no official name for this series of compounds is available, so for lack of a better name, A-series nerve agents) designated A-230, A-232, and A-234, which had been created earlier.

The compounds, A-230, A-232, and A-234 have recently been cited in the literature with reference to their chemical structure but with no analytical data. They are listed in Chapter 5, but only their chemical structure, chemical name, chemical formula, molecular mass, and CAS number are provided.

The compounds that have been referred to as novichok–#, novichok-5, and novichok-7 are reportedly the binary counterparts for the above mentioned A-series agents.

Very little chemistry is known about the novichoks. However, from the possible synthesis listed in chemical abstracts for the compounds A-230, A-232, and A-234 we can infer what the novichok's may be.

Do bear in mind that most of this is an educated guess and I may not be totally correct.

Figure 5.1. Novichok Compounds.

Table 5.1.

R	R'	Compound	CAS #	A-Series
H	H	Novichok-??	[765-40-2]	A-230
H	CH_3	Novichok-5	[16415-09-1]	A-232
CH_3	CH_3	Novichok-7	[19952-57-9]	A-234

The synthetic route cited in the literature is as follows:

Under subzero temperatures compounds were stable but upon being warmed they underwent ring opening to form one of the A-series agents.

I hope that at least this gives some insight as to the chemistry of the novichok compounds, though its very little and even questionable.

I hope all that use this book find the information useful and practical. Suggestions are always welcomed.

5.1. Nerve Agent – A-230

This series of agents are a new class of nerve agents developed by the former Soviet Union. Very little information is available about them.. It has been reported that this class of agents are five to eight, possibly as much as 10× stronger as VX. They are conjectured to be the unitary nerve agents from the binary novichoks. A-230 maybe the result of novichok-# and novichok-?? combining.

Informational

Designation	A-230
Class	Nerve agent
Type	No data available
Chemical name	[[(2-chloroethoxy)fluorohydroxyphosphinyl]oxy] carbonimidic chloride fluoride
CAS number	[26102-97-6]

Chemical and Physical Properties

Appearance	No data available
Odor	No data available
Chemical formula	$C_3H_4Cl_2F_2NO_3P$
Molecular weight	241.93
Chemical structure	
Melting point	No data available
Boiling point	No data available

Flash point	No data available
Decomposition temperature	No data available
Vapor density	No data available
Liquid density	No data available
Solid density	No data available
Vapor pressure	No data available
Volatility	No data available
Solubility	No data available

Reactivity

Hydrolysis products	No data available
Rate of hydrolysis	No data available
Stability	No data available
Storage stability	No data available
Decomposition	No data available
Polymerization	No data available

Toxicity

LD_{50} *(skin)*	No data available
LCt_{50} *(respiratory)*	No data available
LCt_{50} *(percutaneous)*	No data available
ICt_{50} *(respiratory)*	No data available
ICt_{50} *(percutaneous)*	No data available
Rate of detoxification	No data available
Skin and eye toxicity	No data available
Rate of action	No data available
Overexposure effects	Signs and symptoms are the same regardless of route the poison enters the body (by inhalation, absorption, or ingestion): runny nose; tightness of chest; dimness of vision and miosis (pinpointing of the eye pupils); difficulty in breathing; drooling and excessive sweating; nausea; vomiting; cramps, and involuntary defecation and urination; twitching, jerking, and staggering; and headache, confusion, drowsiness, coma, and convulsion. These signs and symptoms are followed by cessation of breathing and death.

Safety

Protective gloves	Wear Butyl Glove M3 and M4 Norton, Chemical Protective Glove Set.
Eye protection	Wear chemical goggles; use goggles and faceshield for splash hazards.
Other	Wear gloves and lab coat with M9 or M17 mask readily available for general lab work.
Emergency procedures	***Inhalation:*** Hold breath and don respiratory protection mask; administer immediately, in rapid succession, all three Nerve Agent Antidote Kits, Mark I injectors if severe signs of agent exposure appear; use mouth-to-mouth resuscitation when approved mask-bag or oxygen delivery systems are not available, but do not use mouth-to-mouth resuscitation when facial contamination exists; if breathing is difficult, administer oxygen; seek medical attention immediately.
	Eye Contact: Flush eyes immediately with water for 10–15 min, then don a respiratory protective mask. Although miosis may be an early sign of agent exposure, do not administer an injection when miosis is the only sign present; seek medical attention immediately.
	Skin Contact: Don respiratory mask and remove contaminated clothing; wash contaminated skin with copious amounts of soap and water immediately using 10% sodium carbonate solution, or 5% liquid household bleach; rinse well with water to remove decontamination; if local sweating and muscular symptoms occur, administer an intramuscular injection with the MARK I Kit; seek medical attention immediately.
	Ingestion: Do not induce vomiting; first symptoms are likely to be gastrointestinal; administer immediately 2-mg intramuscular injection of the MARK I Kit auto injectors; seek medical attention immediately.

Military Significant Information

Field protection	Protective mask and protective clothing.
Decontamination	HTH, STB slurries; household bleach; DS2 solution; hot, soapy water. Decontaminate liquid agent on the skin with the M258A1, M258, or M291 skin decon kit. Decontaminate individual equipment with the M280 individual equipment decontamination kit.
Persistency	Depends upon munitions used and the weather. Heavily splashed liquid persists for long periods of time under average weather conditions.
Use	Delayed casualty agent.

Reactivity

Hydrolysis products	No data available
Rate of hydrolysis	No data available
Stability	No data available
Storage stability	No data available
Decomposition	No data available
Polymerization	No data available

Toxicity

LD_{50} *(skin)*	No data available
LCt_{50} *(respiratory)*	No data available
LCt_{50} *(percutaneous)*	No data available
ICt_{50} *(respiratory)*	No data available
ICt_{50} *(percutaneous)*	No data available
Rate of detoxification	No data available
Skin and eye toxicity	No data available
Rate of action	No data available
Overexposure effects	Signs and symptoms are the same regardless of route the poison enters the body (by inhalation, absorption, or ingestion): runny nose; tightness of chest; dimness of vision and miosis (pinpointing of the eye pupils); difficulty in breathing; drooling and excessive sweating; nausea; vomiting; cramps, and involuntary defecation and urination; twitching, jerking, and staggering; and headache, confusion, drowsiness, coma, and convulsion. These signs and symptoms are followed by cessation of breathing and death.

Safety

Protective gloves	Wear Butyl Glove M3 and M4 Norton, Chemical Protective Glove Set.
Eye protection	Wear chemical goggles; use goggles and faceshield for splash hazards.
Other	Wear gloves and lab coat with M9 or M17 mask readily available for general lab work.
Emergency procedures	**Inhalation:** Hold breath and don respiratory protection mask; administer immediately, in rapid succession, all three Nerve Agent Antidote Kits, Mark I injectors if severe signs of agent exposure appear; use mouth-to-mouth resuscitation when approved mask-bag or oxygen delivery systems are

not available, but do not use mouth-to-mouth resuscitation when facial contamination exists; if breathing is difficult, administer oxygen; seek medical attention immediately.

Eye Contact: Flush eyes immediately with water for 10–15 min, then don a respiratory protective mask. Although miosis may be an early sign of agent exposure, do not administer an injection when miosis is the only sign present; seek medical attention immediately.

Skin Contact: Don respiratory mask and remove contaminated clothing; wash contaminated skin with copious amounts of soap and water immediately using 10% sodium carbonate solution, or 5% liquid household bleach; rinse well with water to remove decontamination; if local sweating and muscular symptoms occur, administer an intramuscular injection with the MARK I Kit; seek medical attention immediately.

Ingestion: Do not induce vomiting; first symptoms are likely to be gastrointestinal; administer immediately 2-mg intramuscular injection of the MARK I Kit auto injectors; seek medical attention immediately.

Military Significant Information

Field protection	Protective mask and protective clothing.
Decontamination	HTH, STB slurries; household bleach; DS2 solution; hot, soapy water. Decontaminate liquid agent on the skin with the M258A1, M258, or M291 skin decon kit. Decontaminate individual equipment with the M280 individual equipment decontamination kit.
Persistency	Depends upon munitions used and the weather. Heavily splashed liquid persists for long periods of time under average weather conditions.
Use	Delayed casualty agent.

5.3. Nerve Agent – A-234

This series of agents are a new class of nerve agents developed by the former Soviet Union. Very little information is available about them. It has been reported that this class of agents are 5–8, possibly as much as 10× stronger as VX. They are conjectured to be the unitary nerve agents from the binary novichoks. A-234 maybe the result of novichok-# and novichok-7 combining.

Informational

Designation	A-234
Class	Nerve agent

Type	No data available
Chemical name	[[(2-chloro-1-methylpropoxy)fluorohydroxy-phosphinyl]oxy]carbonimidic chloride fluoride
CAS number	[26102-99-8]

Chemical and Physical Properties

Appearance	No data available
Odor	No data available
Chemical formula	$C_5H_8Cl_2F_2NO_3P$
Molecular weight	270.00
Chemical structure	

$$\underset{\text{O}}{\overset{\displaystyle \overset{\text{CH}_3}{|} \quad \overset{\text{CH}_3}{|} \qquad \overset{\text{F}}{|} \qquad\quad \overset{\text{Cl}}{|}}{\text{Cl}-\text{CH}-\text{CH}-\text{O}-\underset{\|}{\text{P}}-\text{O}-\text{N}=\text{C}-\text{F}}}$$

Melting point	No data available
Boiling point	No data available
Flash point	No data available
Decomposition temperature	No data available
Vapor density	No data available
Liquid density	No data available
Solid density	No data available
Vapor pressure	No data available
Volatility	No data available
Solubility	No data available

Reactivity

Hydrolysis products	No data available
Rate of hydrolysis	No data available
Stability	No data available
Storage stability	No data available
Decomposition	No data available
Polymerization	No data available

Toxicity

LD$_{50}$ (skin)	No data available
LCt$_{50}$ (respiratory)	No data available
LCt$_{50}$ (percutaneous)	No data available
ICt$_{50}$ (respiratory)	No data available
ICt$_{50}$ (percutaneous)	No data available
Rate of detoxification	No data available
Skin and eye toxicity	No data available
Rate of action	No data available
Overexposure effects	Signs and symptoms are the same regardless of route the poison enters the body (by inhalation, absorption, or ingestion): runny nose; tightness of chest; dimness of vision and miosis (pinpointing of the eye pupils); difficulty in breathing; drooling and excessive sweating; nausea; vomiting; cramps, and involuntary defecation and urination; twitching, jerking, and staggering; and headache, confusion, drowsiness, coma, and convulsion. These signs and symptoms are followed by cessation of breathing and death.

Safety

Protective gloves	Wear Butyl Glove M3 and M4 Norton, Chemical Protective Glove Set.
Eye protection	Wear chemical goggles; use goggles and faceshield for splash hazards.
Other	Wear gloves and lab coat with M9 or M17 mask readily available for general lab work.
Emergency procedures	***Inhalation:*** Hold breath and don respiratory protection mask; administer immediately, in rapid succession, all three Nerve Agent Antidote Kits, Mark I injectors if severe signs of agent exposure appear; use mouth-to-mouth resuscitation when approved mask-bag or oxygen delivery systems are not available, but do not use mouth-to-mouth resuscitation when facial contamination exists; if breathing is difficult, administer oxygen; seek medical attention immediately.

Eye Contact: Flush eyes immediately with water for 10–15 min, then don a respiratory protective mask. Although miosis may be an early sign of agent exposure, do not administer an injection when miosis is the only sign present; seek medical attention immediately.

Skin Contact: Don respiratory mask and remove contaminated clothing; wash contaminated skin with copious amounts of soap and water immediately using 10% sodium |

carbonate solution, or 5% liquid household bleach; rinse well with water to remove decontamination; if local sweating and muscular symptoms occur, administer an intramuscular injection with the MARK I Kit; seek medical attention immediately.

Ingestion: Do not induce vomiting; first symptoms are likely to be gastrointestinal; administer immediately 2-mg intramuscular injection of the MARK I Kit auto injectors; seek medical attention immediately.

Military Significant Information

Field protection	Protective mask and protective clothing.
Decontamination	HTH, STB slurries; household bleach; DS2 solution; hot, soapy water. Decontaminate liquid agent on the skin with the M258A1, M258, or M291 skin decon kit. Decontaminate individual equipment with the M280 individual equipment decontamination kit.
Persistency	Depends upon munitions used and the weather. Heavily splashed liquid persists for long periods of time under average weather conditions.
Use	Delayed casualty agent.

5.4. Amiton – VG

In the 1950s, organophosphates continued to attract interest as pesticides (as well as chemical warfare agents). In 1952, a chemist at the Plant Protection Laboratories of the British firm Imperial Chemical Industries was investigating a class of organophosphate compounds (organophosphate esters of substituted aminoethanethiols). Like earlier investigations of organophosphate, they were found that they were quite effective pesticides. In 1954, ICI put one of them on the market under the trade name Amiton. It was subsequently withdrawn, as it was too toxic for safe use. The toxicity did not go unnoticed, and some of the more toxic materials had, in fact been seen to the British facility at Porton Down for evaluation. After the evaluation was complete, several members of this class of compounds would become a new group of nerve agents, the V-agents. (Depending on who you talk to, the V stands for *V*ictory, *V*enomous, or *V*iscous.)

Informational

Designation	VG
Class	Nerve agent
Type	B – persistent
Chemical name	*O,O*-Diethyl *S*-(2-diethylaminoethyl) thiophosphate
CAS number	[78-53-5]

Chemical and Physical Properties

Appearance	Colorless liquid
Odor	No data available
Chemical formula	$C_{10}H_{24}NO_3PS$
Molecular weight	269.3

Chemical structure

$$CH_3CH_2 \backslash\ N-CH_2CH_2-S-\underset{\underset{OCH_2CH_3}{|}}{\overset{\overset{O}{\|}}{P}}-OCH_2CH_3$$
$$CH_3CH_2 /$$

Melting point	No data available
Boiling point	315°C
Flash point	144°C
Decomposition temperature	No data available
Vapor density	No data available
Liquid density	1.048 g/cm^3 at 25°C
Solid density	No data available
Vapor pressure	0.00054 mmHg at 25°C
Volatility	No data available
Solubility	Highly soluble in water and organic solvents

Reactivity

Hydrolysis products	Ethyl ethoxyphosphonic acid, 2-(Diethylamino) ethanethiol, and *P,P*-diethyl diethoxydiphosphonate
Rate of hydrolysis	Slow
Stability	No data available
Storage stability	No data available
Decomposition	In general, 1,1,4,4 tetraethylpiperazinium salts are formed
Polymerization	No data available

Toxicity

LD$_{50}$ (skin)	No data available
LCt$_{50}$ (respiratory)	No data available
LCt$_{50}$ (percutaneous)	No data available
ICt$_{50}$ (respiratory)	No data available

ICt$_{50}$ (percutaneous)	No data available
Rate of detoxification	No data available
Skin and eye toxicity	No data available
Rate of action	Rapid
Overexposure effects	Excessive salivation, sweating, rhinorrhea, and tearing. Muscle twitching, weakness, tremor, incoordination. Headache, dizziness, nausea, vomiting, abdominal cramps, diarrhea. Respiratory depression, tightness in chest, wheezing, productive cough, fluid in lungs. Pinpoint pupils, sometimes with blurred or dark vision. Severe cases: seizures, incontinence, respiratory depression, loss of consciousness. Cholinesterase inhibition.

Safety

Protective gloves	Wear Butyl Glove M3 and M4 Norton, Chemical Protective Glove Set.
Eye protection	Wear chemical goggles; use goggles and faceshield for splash hazards.
Other	Wear gloves and lab coat with M9 or M17 mask readily available for general lab work.
Emergency procedures	**Inhalation:** Move victims to fresh air. Emergency personnel should avoid self-exposure to amiton. Evaluate vital signs including pulse and respiratory rate, and note any trauma. If no pulse is detected, provide CPR. If not breathing, provide artificial respiration. If breathing is labored, administer 100% humidified oxygen or other respiratory support. Obtain authorization and/or further instructions from the local hospital for administration of an antidote or performance of other invasive procedures. Transport to a health care facility.
	Eye/Skin Contact: Remove victims from exposure. Emergency personnel should avoid self-exposure to amiton. Evaluate vital signs including pulse and respiratory rate, and note any trauma. If no pulse is detected, provide CPR. If not breathing, provide artificial respiration. If breathing is labored, administer 100% humidified oxygen or other respiratory support. Remove contaminated clothing as soon as possible. If eye exposure has occurred, eyes must be flushed with lukewarm water for at least 15 min. Wash exposed skin areas three times with soap and water. Obtain authorization and/or further instructions from the local hospital for administration of an antidote or performance of other invasive procedures. Transport to a health care facility.
	Ingestion: Evaluate vital signs including pulse and respiratory rate, and note any trauma. If no pulse is detected, provide CPR. If not breathing, provide artificial respiration.

If breathing is labored, administer 100% humidified oxygen or other respiratory support. Obtain authorization and/or further instructions from the local hospital for administration of an antidote or performance of other invasive procedures. Vomiting may be induced with syrup of Ipecac. If elapsed time since ingestion of amiton is unknown or suspected to be greater than 30 min, do not induce vomiting. Ipecac should not be administered to children under 6 months of age.

Warning: Ingestion of amiton may result in sudden onset of seizures or loss of consciousness. Syrup of Ipecac should be administered only if victims are alert, have an active gag-reflex, and show no signs of impending seizure or coma The following dosages of Ipecac are recommended: children up to1 year old, 10 mL (1/3 oz); children 1–12 years old, 15 mL (1/2 oz); adults, 30 mL (1 oz). Ambulate (walk) the victims and give large quantities of water. If vomiting has not occurred after 15 min, Ipecac may be readministered. Continue to ambulate and give water to the victims. If vomiting has not occurred within 15 min after second administration of Ipecac, administer activated charcoal. Activated charcoal may be administered if victims are conscious and alert. Use 15–30 g (½ to 1 oz) for children, 50–100 g (1 ¾ to 3 ½ oz) for adults, with 125–250 mL (½ to 1 cup) of water. Promote excretion by administering a saline cathartic or sorbitol to conscious and alert victims. Children require 15–30 g (½ to 1 oz) of cathartic; 50–100 g (1 ¾ to 3 ½ oz) is recommended for adults. Transport to a health care facility.

Military Significant Information

Field protection	Protective mask and protective clothing.
Decontamination	HTH, STB slurries; household bleach; DS2 solution; hot, soapy water. Decontaminate liquid agent on the skin with the M258A1, M258, or M291 skin decon kit. Decontaminate individual equipment with the M280 individual equipment decontamination kit.
Persistency	Depends upon munitions used and the weather. Heavily splashed liquid persists for long periods of time under average weather conditions. In very cold weather VG can persist for months.
Use	Delayed casualty agent.

5.5. Cyclosarin – GF

The agent GF is a fluoride-containing organophosphate. It is a potential nerve agent. It is a slightly volatile liquid that is almost insoluble in water. It enters the body primarily through the respiratory tract but is also highly toxic through the skin and digestive tract. It is a strong cholinesterase inhibitor. Toxicity information reports LD_{50} values in mice from 16 to 400 µg/kg, compared to LD_{50} of 200 µg/kg for Sarin. It is ~20× more persistent that Sarin.

Informational

Designation	GF
Class	Nerve agent
Type	A – nonpersistent
Chemical name	Methylphosphonofluoridic acid, phenyl ester
CAS number	[329-99-7]

Chemical and Physical Properties

Appearance	Clear colorless liquid
Odor	Sweet, musty
Chemical formula	$C_7H_{14}FO_2P$
Molecular weight	180.2

Chemical structure

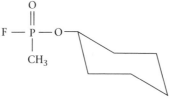

Melting point	–30°C
Boiling point	239°C
Flash point	94°C
Decomposition temperature	No data available
Vapor density	6.2 (air = 1)
Liquid density	1.13 g/cm³ at 20°C
Solid density	No data available
Vapor pressure	0.044 mmHg at 20°C
Volatility	438 mg/m³ at 20°C
	581 mg/m³ at 25°C
Solubility	Almost entirely insoluble in water, 0.37% at 20°C.

Reactivity

Hydrolysis products	GF's hydrolysis products, hydrofluoric and cyclohexyl-methylphosphonic acids, are less toxic than their parent compound.
Rate of hydrolysis	Very stable; hydrolyzes only when heated or in the presence of alkalies
Stability	No data available
Storage stability	Reasonably stable in steel at normal temperatures
Decomposition	No data available
Polymerization	No data available

Toxicity

LD_{50} *(skin)*	16–400 µg/kg (mice)
LCt_{50} *(respiratory)*	35 mg-min/m^3
LCt_{50} *(percutaneous)*	No data available
ICt_{50} *(respiratory)*	No data available
ICt_{50} *(percutaneous)*	No data available
Rate of detoxification	No data available
Skin and eye toxicity	Very high
Rate of action	Rapid
Overexposure effects	No data available

Safety

Protective gloves	Wear Butyl Glove M3 and M4 Norton, Chemical Protective Glove Set.
Eye protection	Wear chemical goggles; use goggles and faceshield for splash hazards.
Other	Wear gloves and lab coat with M9 or M17 mask readily available for general lab work.
Emergency procedures	**Inhalation:** Hold breath and don respiratory protection mask; if severe signs of agent exposure appear, administer immediately, in rapid succession, all three Nerve Agent Antidote Kits, Mark I injectors; use mouth-to-mouth resuscitation when approved mask-bag or oxygen delivery systems are not available; do not use mouth-to-mouth resuscitation when facial contamination exists; administer oxygen if breathing is difficult; seek medical attention immediately.

Eye Contact: Flush eyes immediately with water for 10–15 min then don a respiratory protective mask. Although miosis may be an early sign of agent exposure, do not administer an injection when miosis is the only sign present; seek medical attention immediately.

Skin Contact: Don respiratory mask and remove contaminated clothing; wash contaminated skin with copious amounts of soap and water immediately using 10% sodium carbonate solution, or 5% liquid household bleach; rinse well with water to remove decontamination; administer an intramuscular injection with the Mark I Kit if local sweating and muscular symptoms occur; seek medical attention immediately.

Ingestion: Do not induce vomiting; first symptoms are likely to be gastrointestinal; administer immediately 2-mg intramuscular injection of the MARK I Kit auto injectors; seek medical attention immediately.

Military Significant Information

Field protection	Protective mask and protective clothing.
Decontamination	Flush eyes with water immediately. Use, the M258A1, M258, or M291 skin decontamination kit for liquid agent on the skin. Decontaminate individual equipment with the M280 individual equipment decontamination kit. Calcium hypochlorite (HTH), supertropical bleach (STB), household bleach, caustic soda, dilute alkali solutions, or decontaminating solution number 2 (DS2) are effective on equipment. Use steam and ammonia or hot, soapy water in a confined area.
Persistency	GF is about as persistent as GA. GF evaporates about 20× more slowly than water. Heavily splashed liquid persists 1–2 days under average weather conditions.
Use	Quick-acting casualty agent.

5.6. Edemo – VM

VM also called Edemo is a "V-series" nerve agent closely related to the better-known *VX nerve gas*. Like most of the agents in the V-series (with the exception of VX), VM has not been studied extensively studied. Little known about this compound other than its chemical formula. Since it is structurally very similar to VX it can be assumed that most properties will be similar also.

Informational

Designation	VM
Class	Nerve agent

Type	B – persistent
Chemical name	O-Ethyl S-(2-diethylaminoethyl) methylthio phosphonate
CAS number	[21770-86-5]

Chemical and Physical Properties

Appearance	Colorless to dark yellow oily liquid
Odor	Odorless
Chemical formula	$C_9H_{22}NO_2PS$
Molecular weight	239.3
Chemical structure	

$$CH_3CH_2 \diagdown \quad \quad \quad \quad \quad O$$
$$N-CH_2CH_2-S-\overset{\|}{\underset{\underset{CH_3}{|}}{P}}-OCH_2CH_3$$
$$CH_3CH_2 \diagup$$

Melting point	–50°C
Boiling point	~293°C
Flash point	262.8°C
Decomposition temperature	No data available
Vapor density	8.3 (air = 1)
Liquid density	1.03 g/cm^3 at 25°C
Solid density	No data available
Vapor pressure	0.002 mmHg at 25°C
Volatility	27 mg/m^3 at 25°C
Solubility	Miscible with water

Reactivity

Hydrolysis products	Ethyl methylphosphonic acid, 2-(Diethylamino) ethanethiol, and P,P-diethyl dimethyldiphosphonate
Rate of hydrolysis	Slow
Stability	Persistent; relatively stable at room temperature.
Storage stability	No data available
Decomposition	No data available. However, due to the similarity to VX, decomposition products should be similar.
Polymerization	Will not occur

Toxicity

LD$_{50}$ (skin)	No data available
LCt$_{50}$ (respiratory)	No data available
LCt$_{50}$ (percutaneous)	No data available
ICt$_{50}$ (respiratory)	No data available
ICt$_{50}$ (percutaneous)	No data available
Rate of detoxification	Low; essentially cumulative
Skin and eye toxicity	Extremely toxic by skin and eye absorption. Liquid does not injure the skin or eye but penetrates rapidly. Immediate decontamination of the smallest drop is essential.
Rate of action	Very rapid. Death usually occurs within 15 min after absorption of fatal dosage.
Overexposure effects	Signs and symptoms of overexposure may occur within minutes or hours depending upon dose. They include: miosis (constriction of pupils) and visual effects, headache and pressure sensation, runny nose and nasal congestion, salivation, tightness in the chest, nausea, vomiting, giddiness, anxiety, difficulty in thinking, difficulty sleeping, nightmares, muscle twitches, tremors, weakness, abdominal cramps, diarrhea, involuntary urination and defecation. Signs of severe exposure can progress to convulsions and respiratory failure.

Safety

Protective gloves	Wear Butyl Glove M3 and M4 Norton, Chemical Protective Glove Set.
Eye protection	Wear chemical goggles; use goggles and faceshield for splash hazards.
Other	Wear gloves and lab coat with M9 or M17 mask readily available for general lab work.
Emergency procedures	**Inhalation:** Hold breath and don respiratory protection mask; administer immediately, in rapid succession, all three Nerve Agent Antidote Kits, Mark I injectors if severe signs of agent exposure appear; use mouth-to-mouth resuscitation when approved mask-bag or oxygen delivery systems are not available, but do not use mouth-to-mouth resuscitation when facial contamination exists; if breathing is difficult, administer oxygen; seek medical attention immediately.

Eye Contact: Flush eyes immediately with water for 10–15 min, then don a respiratory protective mask. Although miosis may be an early sign of agent exposure, do not administer an injection when miosis is the only sign present; seek medical attention immediately. |

Skin Contact: Don respiratory mask and remove contaminated clothing; wash contaminated skin with copious amounts of soap and water immediately using 10% sodium carbonate solution, or 5% liquid household bleach; rinse well with water to remove decontamination; if local sweating and muscular symptoms occur, administer an intramuscular injection with the MARK I Kit; seek medical attention immediately.

Ingestion: Do not induce vomiting; first symptoms are likely to be gastrointestinal; administer immediately 2-mg intramuscular injection of the MARK I Kit auto injectors; seek medical attention immediately.

Military Significant Information

Field protection	Protective mask and protective clothing.
Decontamination	HTH, STB slurries; household bleach; DS2 solution; hot, soapy water. Decontaminate liquid agent on the skin with the M258A1, M258, or M291 skin decon kit. Decontaminate individual equipment with the M280 individual equipment decontamination kit.
Persistency	Depends upon munitions used and the weather. Heavily splashed liquid persists for long periods of time under average weather conditions. In very cold weather VM can persist for months. VM is estimated to be slower in evaporating than GB.
Use	Delayed casualty agent.

5.7. Ethyl Sarin – GE

Although GE is not a well-known nerve agent. Some experimental work was done at the Defence Science and Technology Laboratory at Porton Down, United Kingdom, during the service volunteer program (1939–1989). Since GE is a structural analog to GB, most information can be inferred form the data pertaining to GB.

Informational

Designation	GE
Class	Nerve agent
Type	A – nonpersistent
Chemical name	Isopropyl-ethylphosphonofluoridate
CAS number	[1189-87-3]

Chemical and Physical Properties

Appearance	No data available
Odor	No data available
Chemical formula	$C_5H_{12}FO_2P$
Molecular weight	154.12

Chemical structure

$$F - \overset{\overset{\displaystyle O}{\|}}{\underset{\displaystyle CH_2CH_3}{P}} - O - CH(CH_3)_2$$

Melting point	No data available
Boiling point	170°C
Flash point	56.7°C
Decomposition temperature	No data available
Vapor density	No data available
Liquid density	1.05 g/cm³ at 25°C
Solid density	No data available
Vapor pressure	1.97 mmHg at 25°C
Volatility	No data available
Solubility	Soluble in all organic solvents, including alcohols, gasoline, oils, and fats. GE is miscible with water.

Reactivity

Hydrolysis products	Hydrogen fluoride under acid conditions; isopropyl alcohol and polymers under alkaline conditions.
Rate of hydrolysis	Variable with pH
Stability	Similar to GB
Storage stability	Similar to GB
Decomposition	No data available
Polymerization	Similar to GB

Toxicity

LD_{50} *(skin)*	Similar to GB
LCt_{50} *(respiratory)*	Similar to GB
LCt_{50} *(percutaneous)*	No data available
ICt_{50} *(respiratory)*	Similar to GB

ICt$_{50}$ (percutaneous)   No data available

Rate of detoxification   Low; essentially cumulative

Skin and eye toxicity   **Eyes:** Very high toxicity; much greater through eyes than through skin. Vapor causes pupils of the eyes to constrict; vision becomes difficult in dim light.

Skin: Liquid does not injure skin but penetrates it rapidly. Immediate decontamination of smallest drop is essential. Vapor penetrates skin also.

Rate of action   Very rapid; death usually occurs within 15 min after absorption of fatal dose.

Overexposure effects   Signs and symptoms are the same regardless of route the poison enters the body (by inhalation, absorption, or ingestion): runny nose; tightness of chest; dimness of vision and miosis (pinpointing of the eye pupils); difficulty in breathing; drooling and excessive sweating; nausea; vomiting; cramps, and involuntary defecation and urination; twitching, jerking, and staggering; and headache, confusion, drowsiness coma, and convulsion. These signs and symptoms are followed by cessation of breathing and death.

Safety

Protective gloves   Wear Butyl Glove M3 and M4 Norton, Chemical Protective Glove Set.

Eye protection   Wear chemical goggles; use goggles and faceshield for splash hazards.

Other   Wear gloves and lab coat with M9 or M17 mask readily available for general lab work.

Emergency procedures   **Inhalation:** Hold breath and don respiratory protection mask; if severe signs of agent exposure appear, administer immediately, in rapid succession, all three Nerve Agent Antidote Kits, Mark I injectors; use mouth-to-mouth resuscitation when approved mask-bag or oxygen delivery systems are not available; do not use mouth-to-mouth resuscitation when facial contamination exists; administer oxygen if breathing is difficult; seek medical attention immediately.

Eye Contact: Flush eyes immediately with water for 10–15 min then don a respiratory protective mask. Although miosis may be an early sign of agent exposure, do not administer an injection when miosis is the only sign present; seek medical attention immediately.

Skin Contact: Don respiratory mask and remove contaminated clothing; wash contaminated skin with copious amounts of soap and water immediately using 10% sodium carbonate solution, or 5% liquid household bleach; rinse well with water to remove decontamination; administer an intramuscular injection with the Mark I Kit if local sweating and muscular symptoms occur; seek medical attention immediately.

Ingestion: Do not induce vomiting; first symptoms are likely to be gastrointestinal; administer immediately 2-mg intramuscular injection of the MARK I Kit auto injectors; seek medical attention immediately.

Military Significant Information

Field protection	Protective mask and protective clothing. Clothing gives off G-agents for about 30 min after contact with vapor; consider this fact before unmasking. Immediately remove all liquid from clothing.
Decontamination	Flush eyes with water immediately. Use, the M258A1, M258, or M291 skin decontamination kit for liquid agent on the skin. Decontaminate individual equipment with the M280 individual equipment decontamination kit. Calcium hypochlorite (HTH), supertropical bleach (STB), household bleach, caustic soda, dilute alkali solutions, or decontaminating solution number 2 (DS2) are effective on equipment. Use steam and ammonia or hot, soapy water in a confined area.
Persistency	Evaporates at approximately the same rate as water; depends upon munitions used and the weather.
Use	Quick-acting casualty agent.

5.8. Nerve Agent – GV

The GV (2-dialkylaminoalkyl *N,N*-dialkylphosphonamidofluoridate) nerve agent has a toxicity intermediate to G- and V-type nerve agents. This class of compounds can be described in general as 2-dialkylaminoalkyl-(dialkylamido)-fluorophosphates. These are structurally similar with the groups of G-compounds (i.e., sarin, soman, tabun) and V-compounds (i.e., VX and others). These chemicals were designated as GV compounds. The volatility of GV compounds is between VX and GB (sarin) and therefore these agents are effective when penetrating through uniforms. Intoxication with this compound is practically the same as is observed for nerve agents. Treatment with reactivators and atropine is difficult because the complex formed with cholinesterase is irreversible. The lack of reactivation is different from that observed for GD (soman) (aging, dealkylation) and may be caused by steric hindrance in the cavity of cholinesterase.

Informational

Designation	GV
Class	Nerve agent

Type	A – nonpersistent
Chemical name	2-(dimethylamino)ethyl-dimethylphosphoramido fluoridate
CAS number	[141102-74-1]

Chemical and Physical Properties

Appearance	Colorless to white semisolid
Odor	No data available
Chemical formula	$C_6H_{16}FN_2O_2P$
Molecular weight	198.18

Chemical structure

$$F-\overset{\overset{\displaystyle O}{\|}}{\underset{\underset{\displaystyle CH_3 \quad CH_3}{\diagup \quad \diagdown}}{\underset{\displaystyle N}{|}}}P-O-CH_2CH_2-N\overset{\diagup CH_3}{\diagdown CH_3}$$

Melting point	–110°C
Boiling point	226°C
Flash point	83°C
Decomposition temperature	No data available
Vapor density	6.8 (air = 1)
Liquid density	1.11 at 25°C
Solid density	No data available
Vapor pressure	0.049 mmHg at 25°C
Volatility	527 mg/m³ at 25°C
Solubility	No data available

Reactivity

Hydrolysis products	Hydrogen fluoride, amines and complex organo phosphates.
Rate of hydrolysis	No data available
Stability	Unstable undergoes spontaneous isomerization.
Storage stability	No data available
Decomposition	Slow in water, pH dependent.
Polymerization	Undergoes spontaneous isomerization to dimethylaziri-dinium-dimethylamidofluorophosphate.

Toxicity

LD_{50} *(skin)*	11–1222 µg/kg (rodents)
LCt_{50} *(respiratory)*	No data available
LCt_{50} *(percutaneous)*	No data available
ICt_{50} *(respiratory)*	No data available
ICt_{50} *(percutaneous)*	No data available
Rate of detoxification	Very slow, essentially cumulative
Skin and eye toxicity	No data available
Rate of action	Rapid
Overexposure effects	In contrast to either "G" or "V" series nerve agents, observable signs tend to be more mild and transient. Lethal amounts of vapor exposure cause loss of consciousness and convulsions within 30 s to 2 min of exposure, followed by cessation of breathing and paralysis after several additional minutes. General signs and symptoms include sweating, nausea, vomiting, involuntary urination/defecation, and a feeling of weakness.

Safety

Protective gloves	Wear Butyl Glove M3 and M4 Norton, Chemical Protective Glove Set.
Eye protection	Wear chemical goggles; use goggles and faceshield for splash hazards.
Other	Wear gloves and lab coat with M9 or M17 mask readily available for general lab work.
Emergency procedures	**Inhalation:** Hold breath and don respiratory protection mask; administer immediately, in rapid succession, all three Nerve Agent Antidote Kits, Mark I injectors if severe signs of agent exposure appear; use mouth-to-mouth resuscitation when approved mask-bag or oxygen delivery systems are not available, but do not use mouth-to-mouth resuscitation when facial contamination exists; if breathing is difficult, administer oxygen; seek medical attention immediately.
	Eye Contact: Flush eyes immediately with water for 10–15 min, then don a respiratory protective mask. Although miosis may be an early sign of agent exposure, do not administer an injection when miosis is the only sign present; seek medical attention immediately.
	Skin Contact: Don respiratory mask and remove contaminated clothing; wash contaminated skin with copious amounts of soap and water immediately using 10% sodium carbonate solution, or 5% liquid household bleach; rinse well with water to remove decontamination; if local

sweating and muscular symptoms occur, administer an intramuscular injection with the MARK I Kit; seek medical attention immediately.

Ingestion: Do not induce vomiting; first symptoms are likely to be gastrointestinal; administer immediately 2-mg intramuscular injection of the MARK I Kit auto injectors; seek medical attention immediately.

Military Significant Information

Field protection	Protective mask and protective clothing.
Decontamination	HTH, STB slurries; household bleach; DS2 solution; hot, soapy water. Decontaminate liquid agent on the skin with the M258A1, M258, or M291 skin decon kit. Decontaminate individual equipment with the M280 individual equipment decontamination kit.
Persistency	Significantly more persistent than other "G" series nerve agents but with a significantly greater vapor pressure than the "V" series nerve agents.
Use	Quick-acting casualty agent.

5.9. Nerve Agent – VE

VE is a "V-series" nerve agent closely related to the better-known VX nerve gas. Like most of the agents in the V-series (with the exception of VX), VE has not been extensively studied outside of military science. Little is known about this compound other than its chemical formula. It is commonly theorized that the so called "second-generation" V-series agents came from a cold war era Russian chemical weapons development program. They may have been developed sometime between 1950 and 1990. They have similar lethal dose levels to VX (between 10 and 50 mg) and have similar symptoms and method of action to other nerve agents that act on cholinesterase, and treatment remains the same, but the window for effectively treating second-generation V-series seizures is shorter. In addition to the standard seizures, some of the second-generation V-series agents are known to cause coma.

Informational

Designation	VE
Class	Nerve agent
Type	B – persistent
Chemical name	O-ethyl-S-diethylaminoethylphosonothiolate
CAS number	[21738-25-0]

Chemical and Physical Properties

Appearance	No data available
Odor	No data available
Chemical formula	$C_{10}H_{24}NO_2PS$
Molecular weight	253.34

Chemical structure

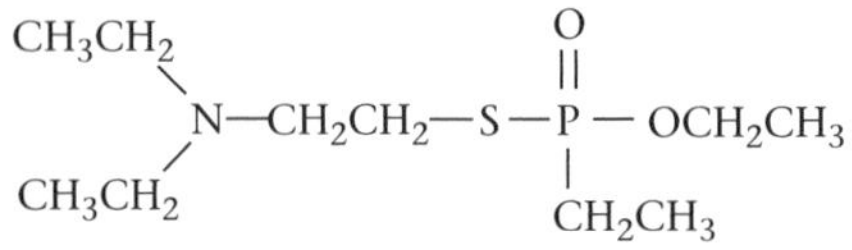

$$CH_3CH_2 \diagdown$$
$$N-CH_2CH_2-S-P-OCH_2CH_3$$
$$CH_3CH_2 \diagup$$

with O double bonded to P and CH_2CH_3 below P

Melting point	No data available
Boiling point	311°C
Flash point	142°C
Decomposition temperature	No data available
Vapor density	No data available
Liquid density	No data available
Solid density	No data available
Vapor pressure	0.000552 mmHg at 25°C
Volatility	No data available
Solubility	No data available

Reactivity

Hydrolysis products	Ethyl ethylphosphonic acid, 2-(diethylamino) ethanethiol, and *P,P*-diethyl diethyldiphosphonate
Rate of hydrolysis	No data available
Stability	No data available
Storage stability	No data available
Decomposition	No data available
Polymerization	No data available

Toxicity

LD$_{50}$ (skin)	No data available
LCt$_{50}$ (respiratory)	No data available
LCt$_{50}$ (percutaneous)	No data available
ICt$_{50}$ (respiratory)	No data available

ICt$_{50}$ (percutaneous)   No data available

Rate of detoxification   Low; essentially cumulative.

Skin and eye toxicity   Extremely toxic by skin and eye absorption. Liquid does not injure the skin or eye but penetrates rapidly. Immediate decontamination of the smallest drop is essential.

Rate of action   Very rapid. Death usually occurs within 15 min after absorption of fatal dosage.

Overexposure effects   Signs and symptoms of overexposure may occur within minutes or hours depending upon dose. They include: miosis (constriction of pupils) and visual effects, headache and pressure sensation, runny nose and nasal congestion, salivation, tightness in the chest, nausea, vomiting, giddiness, anxiety, difficulty in thinking, difficulty sleeping, nightmares, muscle twitches, tremors, weakness, abdominal cramps, diarrhea, involuntary urination and defecation. Signs of severe exposure can progress to convulsions and respiratory failure.

Safety

Protective gloves   Wear Butyl Glove M3 and M4 Norton, Chemical Protective Glove Set.

Eye protection   Wear chemical goggles; use goggles and faceshield for splash hazards.

Other   Wear gloves and lab coat with M9 or M17 mask readily available for general lab work.

Emergency procedures   **Inhalation:** Hold breath and don respiratory protection mask; administer immediately, in rapid succession, all three Nerve Agent Antidote Kits, Mark I injectors if severe signs of agent exposure appear; use mouth-to-mouth resuscitation when approved mask-bag or oxygen delivery systems are not available, but do not use mouth-to-mouth resuscitation when facial contamination exists; if breathing is difficult, administer oxygen; seek medical attention immediately.

Eye Contact: Flush eyes immediately with water for 10–15 min, then don a respiratory protective mask. Although miosis may be an early sign of agent exposure, do not administer an injection when miosis is the only sign present; seek medical attention immediately.

Skin Contact: Don respiratory mask and remove contaminated clothing; wash contaminated skin with copious amounts of soap and water immediately using 10% sodium carbonate solution, or 5% liquid household bleach; rinse well with water to remove decontamination; if local sweating and muscular symptoms occur, administer an intramuscular injection with the MARK I Kit; seek medical attention immediately.

Ingestion: Do not induce vomiting; first symptoms are likely to be gastrointestinal; administer immediately 2-mg intramuscular injection of the MARK I Kit auto injectors; seek medical attention immediately.

Military Significant Information

Field protection	Protective mask and protective clothing.
Decontamination	HTH, STB slurries; household bleach; DS2 solution; hot, soapy water. Decontaminate liquid agent on the skin with the M258A1, M258, or M291 skin decon kit. Decontaminate individual equipment with the M280 individual equipment decontamination kit.
Persistency	Depends upon munitions used and the weather. Heavily splashed liquid persists for long periods of time under average weather conditions. In very cold weather VE can persist for months.
Use	Delayed casualty agent.

5.10. Nerve Agent – VS

VS is a "V-series" nerve agent closely related to the better-known VX nerve gas. Like most of the agents in the V-series (with the exception of VX), VS has not been studied extensively studied. Little known about this compound other than its chemical formula. Since it is structurally very similar to VX it can be assumed that most properties will be similar also.

Informational

Designation	VS
Class	Nerve agent
Type	B – persistent
Chemical name	O-Ethyl-S-2-(diisopropylamino)ethylethylphosphonothiolate
CAS number	[73835-17-3]

Chemical and Physical Properties

Appearance	No data available
Odor	No data available
Chemical formula	$C_{12}H_{28}NO_2PS$

Molecular weight	281.4

Chemical structure

$$(CH_3)_2CH \diagdown \atop (CH_3)_2CH \diagup N-CH_2CH_2-S-\underset{\underset{CH_2CH_3}{\displaystyle |}}{\overset{\overset{\displaystyle O}{\displaystyle \|}}{P}}-OCH_2CH_3$$

Melting point	No data available
Boiling point	335°C
Flash point	156.5°C
Decomposition temperature	No data available
Vapor density	No data available
Liquid density	No data available
Solid density	No data available
Vapor pressure	0.000122 mmHg at 25°C
Volatility	No data available
Solubility	No data available

Reactivity

Hydrolysis products	Diethylphosphonic acid, 2-(Diisopropylamino) ethanethiol, and *P,P*-diethyl diethyldiphosphonate.
Rate of hydrolysis	Slow
Stability	Persistent; relatively stable at room temperature.
Storage stability	No data available
Decomposition	No data available. However, due to the similarity to VX, decomposition products should be similar.
Polymerization	Will not occur

Toxicity

LD$_{50}$ (skin)	No data available
LCt$_{50}$ (respiratory)	No data available
LCt$_{50}$ (percutaneous)	
ICt$_{50}$ (respiratory)	No data available
ICt$_{50}$ (percutaneous)	
Rate of detoxification	Low; essentially cumulative

Skin and eye toxicity	Extremely toxic by skin and eye absorption. Liquid does not injure the skin or eye but penetrates rapidly. Immediate decontamination of the smallest drop is essential.
Rate of action	Very rapid. Death usually occurs within 15 min after absorption of fatal dosage.
Overexposure effects	Signs and symptoms of overexposure may occur within minutes or hours depending upon dose. They include: miosis (constriction of pupils) and visual effects, headache and pressure sensation, runny nose and nasal congestion, salivation, tightness in the chest, nausea, vomiting, giddiness, anxiety, difficulty in thinking, difficulty sleeping, nightmares, muscle twitches, tremors, weakness, abdominal cramps, diarrhea, involuntary urination and defecation. Signs of severe exposure can progress to convulsions and respiratory failure.

Safety

Protective gloves	Wear Butyl Glove M3 and M4 Norton, Chemical Protective Glove Set.
Eye protection	Wear chemical goggles; use goggles and faceshield for splash hazards.
Other	Wear gloves and lab coat with M9 or M17 mask readily available for general lab work.
Emergency procedures	**Inhalation:** Hold breath and don respiratory protection mask; administer immediately, in rapid succession, all three Nerve Agent Antidote Kits, Mark I injectors if severe signs of agent exposure appear; use mouth-to-mouth resuscitation when approved mask-bag or oxygen delivery systems are not available, but do not use mouth-to-mouth resuscitation when facial contamination exists; if breathing is difficult, administer oxygen; seek medical attention immediately.
	Eye Contact: Flush eyes immediately with water for 10–15 min, then don a respiratory protective mask. Although miosis may be an early sign of agent exposure, do not administer an injection when miosis is the only sign present; seek medical attention immediately.
	Skin Contact: Don respiratory mask and remove contaminated clothing; wash contaminated skin with copious amounts of soap and water immediately using 10% sodium carbonate solution, or 5% liquid household bleach; rinse well with water to remove decontamination; if local sweating and muscular symptoms occur, administer an intramuscular injection with the MARK I Kit; seek medical attention immediately.
	Ingestion: Do not induce vomiting; first symptoms are likely to be gastrointestinal; administer immediately 2-mg intramuscular injection of the MARK I Kit auto injectors; seek medical attention immediately.

Military Significant Information

Field protection	Protective mask and protective clothing.
Decontamination	HTH, STB slurries; household bleach; DS2 solution; hot, soapy water. Decontaminate liquid agent on the skin with the M258A1, M258, or M291 skin decon kit. Decontaminate individual equipment with the M280 individual equipment decontamination kit.
Persistency	Depends upon munitions used and the weather. Heavily splashed liquid persists for long periods of time under average weather conditions. In very cold weather VS can persist for an extended period of time.
Use	Delayed casualty agent

5.11. Nerve Agent – VX

VX is a lethal anticholinesterase agent. It is a persistent, nonvolatile agent that is primarily a liquid exposure hazard to the skin or eyes. VX affects the body by blocking the action of the enzyme acetycholinesterase. When this enzyme is blocked, large amounts of the chemical acetylcholine build up at critical places within the nervous system, causing hyperactivity of the body organs stimulated by these nerves. The signs and symptoms of exposure to Nerve agent VX depend upon the *route of exposure* and the *amount of exposure*.

Informational

Designation	VX
Class	Nerve agent
Type	B – persistent
Chemical name	O-ethyl S-(2-diisopropylaminoethyl) methylphosphonothiolate
CAS number	[50782-69-9]

Chemical and Physical Properties

Appearance	Oily liquid that is clear, and tasteless. It is amber colored similar in appearance to motor oil.
Odor	Odorless
Chemical formula	$C_{11}H_{26}NO_2PS$
Molecular weight	267.37

Chemical structure	$(CH_3)_2CH$ $\quad\quad\quad \searrow$ $\quad\quad\quad\quad N - CH_2CH_2 - S - \overset{\displaystyle O}{\overset{\displaystyle \|}{P}} - OCH_2CH_3$ $\quad\quad\quad \nearrow \quad\quad\quad\quad\quad\quad\quad\quad |$ $(CH_3)_2CH \quad\quad\quad\quad\quad\quad\quad\quad CH_3$
Melting point	$< -51°C$ because of dissolved impurities; $-39°C$ calculated.
Boiling point	$298°C$ (calculated) decomposes.
Flash point	$159°C$
Decomposition temperature	Half-life: 36 h at $150°C$; 1.6 h at $200°C$; 4 min at $250°C$; 36 s at $295°C$.
Vapor density	9.2 (air = 1)
Liquid density	1.0083 g/cm^3 at $25°C$
Solid density	No data available
Vapor pressure	0.0007 mmHg at $25°C$
Volatility	10.5 mg/m^3 at $25°C$
Solubility	Miscible with water below $9.4°C$. Slightly soluble in water at room temperature. Soluble in organic solvents.

Reactivity

Hydrolysis products	Ethyl methylphosphonic acid, 2-(Diisopropylamino) ethanethiol, and *P,P*-diethyl dimethyldiphosphonate.
Rate of hydrolysis	Half-life at $25°C$: 100 days at pH 2 or 3; 16 min at pH 13; 1.3 min at pH 14.
Stability	Persistent; relatively stable at room temperature; unstabilized VX of 95% purity decomposed at a rate of 5% a month at $71°C$.
Storage stability	No data available
Decomposition	During basic hydrolysis of VX up to about 10% of the agent if converted to EA2191 (diisopropylaminoethyl methylphosphonothioic acid). Based on the concentration of EA2192 expected to be formed during hydrolysis and its toxicity (1.4 mg/kg dermal in rabbit at 24 h in a 10/90% by wt ethanol/water solution), a Class B poison would result. A large-scale decon procedure, which uses both HTH and NaOH, destroys VX by oxidation and hydrolysis. Typically, the large-scale product contains 0.2–0.4% by wt EA2192 at 24 h. At pH 12, the EA2192 in the large-scale product has a half-life of about 14 days. Thus, a 90-day holding period at pH 12 results in about a 64-fold reduction of EA2192 (6 half-lives). This holding period has been shown to be sufficient to reduce the toxicity of the product below that

of a Class B poison. Other less toxic products are ethyl methylphosphonic acid, methylphosphonic acid, diisopropylaminoethyl mercaptan, diethyl methylphosphonate, and ethanol. A small scale decontamination procedure uses sufficient HTH to oxidize all VX; thus no EA2192 is formed.

Polymerization	Will not occur

Toxicity

LD_{50} *(skin)*	10 mg/person (bare skin)
LCt_{50} *(respiratory)*	30 mg-min/m^3 (mild activity)
LCt_{50} *(percutaneous)*	100 mg-min/m^3 (resting) 6–360 mg-min/m^3 (bare skin) 6–3,600 mg-min/m^3 (clothed)
ICt_{50} *(respiratory)*	24 mg-min/m^3 (mild activity)
ICt_{50} *(percutaneous)*	50 mg-min/m^3 (resting) No data available
Rate of detoxification	Low; essentially cumulative
Skin and eye toxicity	Extremely toxic by skin and eye absorption; about 100× as potent as GB. Liquid does not injure the skin or eye but penetrates rapidly. Immediate decontamination of the smallest drop is essential.
Rate of action	Very rapid. Death usually occurs within 15 min after absorption of fatal dosage.
Overexposure effects	Signs and symptoms of overexposure may occur within minutes or hours depending upon dose. They include: miosis (constriction of pupils) and visual effects, headache and pressure sensation, runny nose and nasal congestion, salivation, tightness in the chest, nausea, vomiting, giddiness, anxiety, difficulty in thinking, difficulty sleeping, nightmares, muscle twitches, tremors, weakness, abdominal cramps, diarrhea, involuntary urination and defecation. Signs of severe exposure can progress to convulsions and respiratory failure.

Safety

Protective gloves	Wear Butyl Glove M3 and M4 Norton, Chemical Protective Glove Set.
Eye protection	Wear chemical goggles; use goggles and faceshield for splash hazards.
Other	Wear gloves and lab coat with M9 or M17 mask readily available for general lab work.

Emergency procedures **Inhalation:** Hold breath and don respiratory protection mask; administer immediately, in rapid succession, all three Nerve Agent Antidote Kits, Mark I injectors if severe signs of agent exposure appear; use mouth-to-mouth resuscitation when approved mask-bag or oxygen delivery systems are not available, but do not use mouth-to-mouth resuscitation when facial contamination exists; if breathing is difficult, administer oxygen; seek medical attention immediately.

Eye Contact: Flush eyes immediately with water for 10–15 min, then don a respiratory protective mask. Although miosis may be an early sign of agent exposure, do not administer an injection when miosis is the only sign present; seek medical attention immediately.

Skin Contact: Don respiratory mask and remove contaminated clothing; wash contaminated skin with copious amounts of soap and water immediately using 10% sodium carbonate solution, or 5% liquid household bleach; rinse well with water to remove decontamination; if local sweating and muscular symptoms occur, administer an intramuscular injection with the MARK I Kit; seek medical attention immediately.

Ingestion: Do not induce vomiting; first symptoms are likely to be gastrointestinal; administer immediately 2-mg intramuscular injection of the MARK I Kit auto injectors; seek medical attention immediately.

Military Significant Information

Field protection   Protective mask and protective clothing.

Decontamination   HTH, STB slurries; household bleach; DS2 solution; hot, soapy water. Decontaminate liquid agent on the skin with the M258A1, M258, or M291 skin decon kit. Decontaminate individual equipment with the M280 individual equipment decontamination kit.

Persistency   Depends upon munitions used and the weather. Heavily splashed liquid persists for long periods of time under average weather conditions. In very cold weather VX can persist for months. VX is calculated to be approximately 1,500× slower in evaporating than GB.

Use   Delayed casualty agent.

5.12. Nerve Agent – Vx

Another V-agent of interest is Vx, called "V sub x." Another designation for Vx is "V-gas." The properties of Vx are similar to those of VX. It is nearly 10× more volatile than VX, but is very persistent in comparison to the G-agents. The molecular weight of Vx is 211.2. Listed values are calculated, information on this agent is limited. The physiological action, protection, and decontaminants for Vx are the same as for VX.

Informational

Designation	Vx
Class	Nerve agent
Type	B – persistent
Chemical name	O-ethyl S-(2-dimethylaminoethyl) methylphosphonothiolate
CAS number	[20820-80-8]

Chemical and Physical Properties

Appearance	Oily liquid that is clear, and tasteless. It is amber colored similar in appearance to motor oil.
Odor	Odorless
Chemical formula	$C_7H_{18}NO_2PS$
Molecular weight	211.2

Chemical structure

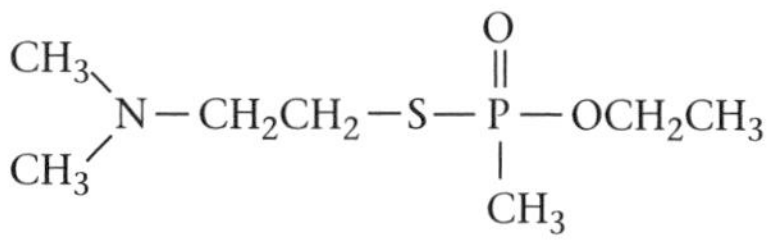

$$CH_3-N-CH_2CH_2-S-\overset{\overset{O}{\|}}{\underset{\underset{CH_3}{|}}{P}}-OCH_2CH_3$$

with CH_3 on the nitrogen

Melting point	No data available.
Boiling point	256°C (calculated)
Flash point	No data available
Decomposition temperature	No data available
Vapor density	7.29 (air = 1)
Liquid density	1.062 g/cm³ at 25°C
Solid density	No data available
Vapor pressure	0.0042 mmHg at 20°C
	0.0066 mmHg at 25°C
Volatility	48 mg/m³ at 20°C
	75 mg/m³ at 25°C
Solubility	Soluble in organic solvents. Slightly soluble in water.

Reactivity

Hydrolysis products	Ethyl methylphosphonic acid, 2-(Dimethylamino) ethanethiol, and P,P-diethyl dimethyldiphosphonate.
Rate of hydrolysis	Slow
Stability	Persistent; relatively stable at room temperature.

Storage stability	No data available
Decomposition	No data available
Polymerization	No data available

Toxicity

LD_{50} *(skin)*	No data available
LCt_{50} *(respiratory),*	No data available
LCt_{50} *(percutaneous)*	No data available
ICt_{50} *(respiratory),*	No data available
ICt_{50} *(percutaneous)*	No data available
Rate of detoxification	Low; essentially cumulative
Skin and eye toxicity	Extremely toxic by skin and eye absorption; about. Liquid does not injure the skin or eye but penetrates rapidly. Immediate decontamination of the smallest drop is essential.
Rate of action	Very rapid. Death usually occurs within 15 min after absorption of fatal dosage.
Overexposure effects	Signs and symptoms of overexposure may occur within minutes or hours depending upon dose. They include: miosis (constriction of pupils) and visual effects, headache and pressure sensation, runny nose and nasal congestion, salivation, tightness in the chest, nausea, vomiting, giddiness, anxiety, difficulty in thinking, difficulty sleeping, nightmares, muscle twitches, tremors, weakness, abdominal cramps, diarrhea, involuntary urination and defecation. Signs of severe exposure can progress to convulsions and respiratory failure.

Safety

Protective gloves	Wear Butyl Glove M3 and M4 Norton, Chemical Protective Glove Set.
Eye protection	Wear chemical goggles; use goggles and faceshield for splash hazards.
Other	Wear gloves and lab coat with M9 or M17 mask readily available for general lab work.
Emergency procedures	**Inhalation:** Hold breath and don respiratory protection mask; administer immediately, in rapid succession, all three Nerve Agent Antidote Kits, Mark I injectors if severe signs of agent exposure appear; use mouth-to-mouth resuscitation when approved mask-bag or oxygen delivery systems are not available, but do not use mouth-to-mouth

resuscitation when facial contamination exists; if breathing is difficult, administer oxygen; seek medical attention immediately.

Eye Contact: Flush eyes immediately with water for 10–15 min, then don a respiratory protective mask. Although miosis may be an early sign of agent exposure, do not administer an injection when miosis is the only sign present; seek medical attention immediately.

Skin Contact: Don respiratory mask and remove contaminated clothing; wash contaminated skin with copious amounts of soap and water immediately using 10% sodium carbonate solution, or 5% liquid household bleach; rinse well with water to remove decontamination; if local sweating and muscular symptoms occur, administer an intramuscular injection with the MARK I Kit; seek medical attention immediately.

Ingestion: Do not induce vomiting; first symptoms are likely to be gastrointestinal; administer immediately 2-mg intramuscular injection of the MARK I Kit auto injectors; seek medical attention immediately.

Military Significant Information

Field protection	Protective mask and protective clothing.
Decontamination	HTH, STB slurries; household bleach; DS2 solution; hot, soapy water. Decontaminate liquid agent on the skin with the M258A1, M258, or M291 skin decontamination kit. Decontaminate individual equipment with the M280 individual equipment decontamination kit.
Persistency	No data available
Use	Delayed casualty agent.

5.13. Russian VX – VR

VR (also has been referred to as Soviet V-gas and Substance 33) and was investigated and developed by the Soviet Union during the 1950s, in a manner which the open literature suggests roughly paralleled the development of VX in the west. It has been suggested that the Soviet investigations were partly guided by knowledge of the molecular formula of VX but lacked information about the structural formula, and so settled on VR, which is a structural isomer of VX. However, it should be noted that there was explosion of interest in this class of compounds during this period, and so it is possible that initial studies had already been undertaken by the Soviets before the United States settled on VX. The greater difficulty of treatment and the possibility that detection might be more difficult could have served as incentives for the choice of VR even if the Soviets were aware of the structural formula for VX.

This is supported by sources that suggest that a pilot production facility for VR was constructed at a facility in Stalingrad (now Volgograd) by the Soviets as early as 1956, 2 years before VX was selected for production by the United States. Research on the agent continued for an extended period as well - in 1974, a Lenin Prize was awarded to researchers who were investigating issues associated with VR.

Informational

Designation	VR
Class	Nerve agent
Type	B – persistent
Chemical name	*O*-isobutyl *S*-(2-diethylaminoethyl)methyl phosphothioate
CAS number	[159939-87-4]

Chemical and Physical Properties

Appearance	Liquid with an "oily" consistency which is colorless when pure.
Odor	No data available
Chemical formula	$C_{11}H_{26}NO_2PS$
Molecular weight	267.38
Chemical structure	CH_3CH_2 and CH_3CH_2 groups attached to $N-CH_2-CH_2-S-P(-OCH_2CH(CH_3)_2)(-CH_3)$ with $P=O$
Melting point	No data available
Boiling point	323°C
Flash point	150°C
Decomposition temperature	No data available
Vapor density	No data available
Liquid density	1.003 g/cm³ at 25°C
Solid density	No data available
Vapor pressure	0.00026 mmHg at 25°C
Volatility	8.9 mg/m³
Solubility	No data available

Reactivity

Hydrolysis products	Isobutyl methylphosphonic acid, 2-(Diethylamino) ethanethiol, and *P,P*-diisobutyl dimethyl diphosphonate.
Rate of hydrolysis	Slow
Stability	Persistent; relatively stable at room temperature.
Storage stability	No data available
Decomposition	No data available. However, due to the similarity to VX, decomposition products should be similar.
Polymerization	Will not occur

Toxicity

LD_{50} *(skin)*	11.3 µg/kg
LCt_{50} *(respiratory),*	No data available
LCt_{50} *(percutaneous)*	No data available
ICt_{50} *(respiratory),*	No data available
ICt_{50} *(percutaneous)*	No data available
Rate of detoxification	Low; essentially cumulative.
Skin and eye toxicity	Extremely toxic by skin and eye absorption. Liquid does not injure the skin or eye but penetrates rapidly. Immediate decontamination of the smallest drop is essential.
Rate of action	Very rapid. Death usually occurs within 15 min after absorption of fatal dosage.
Overexposure effects	Signs and symptoms of overexposure may occur within minutes or hours depending upon dose. They include: miosis (constriction of pupils) and visual effects, headache and pressure sensation, runny nose and nasal congestion, salivation, tightness in the chest, nausea, vomiting, giddiness, anxiety, difficulty in thinking, difficulty sleeping, nightmares, muscle twitches, tremors, weakness, abdominal cramps, diarrhea, involuntary urination and defecation. Signs of severe exposure can progress to convulsions and respiratory failure.

Safety

Protective gloves	Wear Butyl Glove M3 and M4 Norton, Chemical Protective Glove Set.
Eye protection	Wear chemical goggles; use goggles and faceshield for splash hazards.

Other   Wear gloves and lab coat with M9 or M17 mask readily
                        available for general lab work.

Emergency procedures **Inhalation:** Hold breath and don respiratory protection
                        mask; administer immediately, in rapid succession, all three
                        Nerve Agent Antidote Kits, Mark I injectors if severe signs
                        of agent exposure appear; use mouth-to-mouth resuscita-
                        tion when approved mask-bag or oxygen delivery systems
                        are not available, but do not use mouth-to-mouth resuscita-
                        tion when facial contamination exists; if breathing is diffi-
                        cult, administer oxygen; seek medical attention
                        immediately.

                        Eye Contact: Flush eyes immediately with water for 10–15
                        min, then don a respiratory protective mask. Although mio-
                        sis may be an early sign of agent exposure, do not adminis-
                        ter an injection when miosis is the only sign present; seek
                        medical attention immediately.

                        Skin Contact: Don respiratory mask and remove contami-
                        nated clothing; wash contaminated skin with copious
                        amounts of soap and water immediately using 10% sodium
                        carbonate solution, or 5% liquid household bleach; rinse
                        well with water to remove decontamination; if local sweat-
                        ing and muscular symptoms occur, administer an intramus-
                        cular injection with the MARK I Kit; seek medical attention
                        immediately.

                        Ingestion: Do not induce vomiting; first symptoms are
                        likely to be gastrointestinal; administer immediately 2-mg
                        intramuscular injection of the MARK I Kit auto injectors;
                        seek medical attention immediately.

Military Significant Information

Field protection   Protective mask and protective clothing.

Decontamination   HTH, STB slurries; household bleach; DS2 solution; hot,
                        soapy water. Decontaminate liquid agent on the skin with
                        the M258A1, M258, or M291 skin decon kit. Decontaminate
                        individual equipment with the M280 individual equipment
                        decontamination kit.

Persistency   Depends upon munitions used and the weather. Heavily
                        splashed liquid persists for long periods of time under aver-
                        age weather conditions. In very cold weather VR can persist
                        for an extended period of time.

Use   Delayed casualty agent.

5.14. Sarin – GB

GB is a lethal anticholinesterase agent. Its toxic hazard is high for inhalation,
ingestion, and eye/skin exposure. Due to its high volatility, it is mainly an

inhalation threat. Its rate of detoxification is low. Effects of chronic exposures are cumulative. Following a single exposure to GB, daily exposure to concentrations of any nerve agent insufficient to produce symptoms may result in the onset of symptoms after several days. After symptoms subside, increased susceptibility persists for one to several days. The degree of exposure required to produce recurrence, and the severity of these symptoms, depends on duration and time intervals between exposures.

Informational

Designation	GB
Class	Nerve agent
Type	A – nonpersistent
Chemical name	Isopropyl methylphosphonofluoridate
CAS number	[107-44-8]

Chemical and Physical Properties

Appearance	Clear, colorless, and tasteless liquids.
Odor	They are odorless in vapor and pure form.
Chemical formula	$C_4H_{10}FO_2P$
Molecular weight	140.09

Chemical structure

$$F-\overset{\overset{\displaystyle O}{\|}}{\underset{\underset{\displaystyle CH_3}{|}}{P}}-O-\overset{\overset{\displaystyle}{}}{\underset{\underset{\displaystyle CH_3}{|}}{CH}}-CH_3$$

Melting point	–57°C
Boiling point	147°C
Flash point	Does not flash.
Decomposition temperature	Complete decomposition after 2.5 h at 150°C.
Vapor density	4.86 (air = 1)
Liquid density	1.102 g/cm³ at 25°C
Solid density	No data available
Vapor pressure	2.9 mmHg at 25°C
Volatility	4,100 mg/m³ at 0°C 16,091 mg/m³ at 20°C 22,000 mg/m³ at 25°C 29,800 mg/m³ at 30°C

Solubility	Soluble in all organic solvents, including alcohols, gasoline, oils, and fats. GB is miscible with water.

Reactivity

Hydrolysis products	Hydrogen fluoride under acid conditions; isopropyl alcohol and polymers under alkaline conditions.
Rate of hydrolysis	Variable with pH. Half-life 7.5 h at pH 1.8. Very rapidly hydrolyzed in alkaline solutions; half-life 5 h at pH 9. Half-life 30 h in unbuffered solution, 47 h at pH 6.
Stability	Stable when pure, ~20 h.
Storage stability	Fairly stable in steel containers at 65°C. Stability improves with increasing purity. Attacks tin, magnesium, cadmium plated steel, some aluminums; slight attack on copper, brass, lead; practically no attack on 1,020 steel, Inconel and K-monel.
Decomposition	No data available
Polymerization	Under alkaline conditions.

Toxicity

LD_{50} *(skin)*	24 mg/kg
LCt_{50} *(respiratory)*	70–100 mg-min/m^3
LCt_{50} *(percutaneous)*	No data available
ICt_{50} *(respiratory)*	35–75 mg-min/m^3
ICt_{50} *(percutaneous)*	No data available
Rate of detoxification	Low; essentially cumulative
Skin and eye toxicity	*Eyes:* Very high toxicity; much greater through eyes than through skin. Vapor causes pupils of the eyes to constrict; vision becomes difficult in dim light. *Skin:* Lethal dose (LD) is 1.7 g/person. Liquid does not injure skin but penetrates it rapidly. Immediate decontamination of smallest drop is essential. Vapor penetrates skin also.
Rate of action	Very rapid; death usually occurs within 15 min after absorption of fatal dose.
Overexposure effects	Signs and symptoms are the same regardless of route the poison enters the body (by inhalation, absorption, or ingestion): runny nose; tightness of chest; dimness of vision and miosis (pinpointing of the eye pupils); difficulty in breathing; drooling and excessive sweating; nausea; vomiting; cramps, and involuntary defecation and urination; twitching, jerking, and staggering; and headache, confusion, drowsiness, coma, and convulsion. These signs and symptoms are followed by cessation of breathing and death.

Safety

Protective gloves   Wear Butyl Glove M3 and M4 Norton, Chemical Protective
                           Glove Set.

Eye protection   Wear chemical goggles; use goggles and faceshield for splash
                           hazards.

Other   Wear gloves and lab coat with M9 or M17 mask readily
                           available for general lab work.

Emergency procedures   **Inhalation:** Hold breath and don respiratory protection
                           mask; if severe signs of agent exposure appear, administer
                           immediately, in rapid succession, all three Nerve Agent
                           Antidote Kits, Mark I injectors; use mouth-to-mouth resusci-
                           tation when approved mask-bag or oxygen delivery systems
                           are not available; do not use mouth-to-mouth resuscitation
                           when facial contamination exists; administer oxygen if
                           breathing is difficult; seek medical attention immediately.

                           Eye Contact: Flush eyes immediately with water for 10–15
                           min then don a respiratory protective mask. Although mio-
                           sis may be an early sign of agent exposure, do not adminis-
                           ter an injection when miosis is the only sign present; seek
                           medical attention immediately.

                           Skin Contact: Don respiratory mask and remove contami-
                           nated clothing; wash contaminated skin with copious
                           amounts of soap and water immediately using 10% sodium
                           carbonate solution, or 5% liquid household bleach; rinse
                           well with water to remove decontamination; administer an
                           intramuscular injection with the Mark I Kit if local sweating
                           and muscular symptoms occur; seek medical attention
                           immediately.

                           Ingestion: Do not induce vomiting; first symptoms are
                           likely to be gastrointestinal; administer immediately 2-mg
                           intramuscular injection of the MARK I Kit auto injectors;
                           seek medical attention immediately.

Military Significant Information

Field protection   Protective mask and protective clothing. Clothing gives off
                           G-agents for about 30 min after contact with vapor; con-
                           sider this fact before unmasking. Immediately remove all
                           liquid from clothing.

Decontamination   Flush eyes with water immediately. Use, the M258A1, M258,
                           or M291 skin decontamination kit for liquid agent on the
                           skin. Decontaminate individual equipment with the M280
                           individual equipment decontamination kit. HTH, STB,
                           household bleach, caustic soda, dilute alkali solutions, or
                           decontaminating solution number 2 (DS2) are effective on
                           equipment. Use steam and ammonia or hot, soapy water in
                           a confined area.

Persistency	Evaporates at approximately the same rate as water; depends upon munitions used and the weather.
Use	Quick-acting casualty agent.

5.15. Soman – GD

GD is a lethal anticholinesterase agent. Although it is primarily a vapor hazard, its toxic hazard is high for inhalation, ingestion, and eye and skin exposure. Its rate of detoxification in the body is low.

Informational

Designation	GD
Class	Nerve agent
Type	A – nonpersistent.
Chemical name	Pinacolyl methylphosphonofluoridate
CAS number	[96-64-0]

Chemical and Physical Properties

Appearance	Clear, colorless, and tasteless liquids.
Odor	Slight camphor odor and give off a colorless vapor.
Chemical formula	$C_7H_{16}FO_2P$
Molecular weight	182.19
Chemical structure	
Melting point	−42°C (generally solidifies to noncrystalline, glasslike material).
Boiling point	198°C
Flash point	121°C
Decomposition temperature	Stabilized GD decomposes in 200 h at 130°C. Unstablized GD decomposes in 4 h at 130°C.
Vapor density	5.6 (air = 1)
Liquid density	1.02 g/cm³ at 25°C
Solid density	No data available
Vapor pressure	0.40 mmHg at 25°C

Volatility	531 mg/m^3 at 0°C 3,900 mg/m^3 at 25°C 5,570 mg/m^3 at 30°C
Solubility	2.1% at 20°C and 3.4% at 0°C in water. Soluble in sulfur mustard, gasoline, alcohols, fats, and oils.

Reactivity

Hydrolysis products	Essentially hydrogen fluoride.
Rate of hydrolysis	Varies with pH; complete in 5 min in 5% sodium hydroxide (NaOH) solutions. Half-life at pH 6.65 and 25°C is 45 h.
Stability	Less stable than GA or GB.
Storage stability	Stable after storage in steel for 3 months at 65°C. GD corrodes steel at the rate of 1×10^{-5} in./month; ~12 h.
Decomposition	No data available
Polymerization	Will not occur.

Toxicity

LD$_{50}$ (skin)	No data available
LCt$_{50}$ (respiratory)	70–400 mg-min/m^3
LCt$_{50}$ (percutaneous)	10,000 mg-min/m^3 (estimated).
ICt$_{50}$ (respiratory)	35–75 mg-min/m^3
ICt$_{50}$ (percutaneous)	No data available
Rate of detoxification	Low; essentially cumulative.
Skin and eye toxicity	**Eyes:** Very high toxicity; vapor causes pupils of the eyes to constrict; vision becomes difficult in dim light. Toxicity is much greater through eyes than through skin. **Skin:** Extremely toxic through skin absorption. The estimated LD$_{50}$ is 0.35 g/person on bare skin (1.4 g/person in ordinary clothing). Liquid does not injure skin but penetrates it rapidly. Immediate decontamination of smallest drop is essential. Vapor penetrates skin also.
Rate of action	Very rapid; death usually occurs within 15 min after absorption of fatal dose.
Overexposure effects	Signs and symptoms are the same regardless of route the poison enters the body (by inhalation, absorption, or ingestion): runny nose; tightness of chest; dimness of vision and miosis (pinpointing of the eye pupils); difficulty in breathing; drooling and excessive sweating; nausea; vomiting; cramps, and involuntary defecation and urination; twitching, jerking, and staggering; and headache, confusion,

drowsiness, coma, and convulsion. These signs and symptoms are followed by cessation of breathing and death.

Safety

Protective gloves   Wear Butyl Glove M3 and M4 Norton, Chemical Protective Glove Set.

Eye protection   Wear chemical goggles; use goggles and faceshield for splash hazards.

Other   Wear gloves and lab coat with M9 or M17 mask readily available for general lab work.

Emergency procedures   **Inhalation:** Hold breath and don respiratory protection mask; if severe signs of agent exposure appear, administer immediately, in rapid succession, all three Nerve Agent Antidote Kits, Mark I injectors; use mouth-to-mouth resuscitation when approved mask-bag or oxygen delivery systems are not available, but do not use mouth-to-mouth resuscitation when facial contamination exists; administer oxygen if breathing is difficult; seek medical attention immediately.

Eye Contact: Flush eyes immediately with water for 10–15 min then don a respiratory protective mask. Although miosis may be an early sign of agent exposure, do not administer an injection when miosis is the only sign present; seek medical attention immediately.

Skin Contact: Don respiratory mask and remove contaminated clothing; wash contaminated skin with copious amounts of soap and water immediately using 10% sodium carbonate solution, or 5% liquid household bleach; rinse well with water to remove decontamination; administer an intramuscular injection with the Mark I Kit if local sweating and muscular symptoms occur; seek medical attention immediately.

Ingestion: Do not induce vomiting; first symptoms are likely to be gastrointestinal; administer immediately 2-mg intramuscular injection of the MARK I Kit auto injectors; seek medical attention immediately.

Military Significant Information

Field protection   Protective mask and protective clothing. Clothing gives off G-agents for about 30 min after contact with vapor; consider this fact before unmasking. Immediately remove all liquid from clothing.

Decontamination   Flush eyes with water immediately. Use, the M258A1, M258, or M291 skin decontamination kit for liquid agent on the skin. Decontaminate individual equipment with the M280 individual equipment decontamination kit. HTH, STB, household bleach, caustic soda, dilute alkali solutions, or decontaminating solution number 2 (DS2) are effective on equipment. Use steam and ammonia or hot, soapy water in a confined area.

Persistency   Depends upon munitions used and the weather. Heavily splashed liquid persists 1–2 days under average weather conditions. GD is calculated to evaporate about 4× as slowly as water. Addition of agent thickeners can greatly increase persistency.

Use   Quick-acting casualty agent.

5.16. Tabun – GA

G-type nerve agents are considered to be nonpersistent chemical agents that may present a significant vapor hazard to the respiratory tract, eyes, or skin. GA-type nerve agents affect the body by blocking the action of the enzyme AChE. When this enzyme is blocked, large amounts of the chemical acetylcholine build up at critical places within the nervous system, causing hyperactivity of the muscles and body organs stimulated by these nerves. The signs and symptoms of exposure to GA-type nerve agents depend upon the *route of exposure* and the *amount of exposure*.

Informational

Designation	GA
Class	Nerve agent
Type	A – nonpersistent
Chemical name	Dimethylphosphoramidocyanidic acid, ethyl ester
CAS number	[77-81-6]

Chemical and Physical Properties

Appearance	Clear, colorless, and tasteless liquid.
Odor	Slightly fruity odor
Chemical formula	$C_5H_{11}N_2O_2P$
Molecular weight	162.12

Chemical structure

$$CH_3-N(CH_3)-\overset{\overset{O}{\|}}{P}(CN)-O-CH_2CH_3$$

Melting point	−50°C

Boiling point	248°C
Flash point	78°C
Decomposition temperature	Decomposes completely at 150°C for 3.75 h. GA undergoes considerable decomposition when explosively disseminated.
Vapor density	5.6 (air = 1)
Liquid density	1.07 g/cm^3 at 25°C
Solid density	No data available
Vapor pressure	0.037 mmHg at 20°C 0.006 mmHg at 0°C
Volatility	90 mg/m^3 at 0°C 328 mg/m^3 at 20°C 610 mg/m^3 at 25°C 858 mg/m^3 at 30°C
Solubility	Slightly soluble in water; 9.8% at 25°C. 7.2% at 20°C. Readily soluble in organic solvents, such as alcohols, ethers, gasoline, oils, and fats.

Reactivity

Hydrolysis products	Hydrogen cyanide (HCN) and other products.
Rate of hydrolysis	Slow with water but fairly rapid with strong acids and alkalies; self-buffering at pH 4.5. Autocatalytic below pH 4, because of presence of HCN. Half-life, 8.5 h at pH 7 (20°C); 7 h at pH 4–5. The rate of hydrolysis is increased by the presence of phosphate.
Stability	Stable, ~24 h
Storage stability	GA is stable for several years when stored in steel containers at ordinary temperatures.
Decomposition	Decomposes within 6 months at 60°C; complete decomposition in 3.75 h at 150°C; may produce HCN; oxides of nitrogen, oxides of phosphorus, carbon monoxide, and HCN.
Polymerization	No data available

Toxicity

LD$_{50}$ (skin)	1–1.5 mg/person
LCt$_{50}$ (respiratory)	135–400 mg-min/m^3
LCt$_{50}$ (percutaneous)	No data available
ICt$_{50}$ (respiratory)	300 mg-min/m^3
ICt$_{50}$ (percutaneous)	No data available
Rate of detoxification	Slight but definite

Skin and eye toxicity	***Eyes:*** Very high toxicity; much greater through eyes than skin. Very low concentration of vapor causes pupil of eyes to constrict, resulting in difficulty in seeing in dim light.
	Skin: Very toxic. Decontamination of smallest drop of liquid agent is essential. Liquid penetrates skin readily.
Rate of action	No data available
Overexposure effects	Signs and symptoms are the same regardless of route the poison enters the body (by inhalation, absorption, or ingestion): runny nose; tightness of chest; dimness of vision and miosis (pinpointing of the eye pupils); difficulty in breathing; drooling and excessive sweating; nausea; vomiting; cramps, and involuntary defecation and urination; twitching, jerking, and staggering; and headache, confusion, drowsiness, coma, and convulsion. These signs and symptoms are followed by cessation of breathing and death.

Safety

Protective gloves	Wear Butyl Glove M3 and M4 Norton, Chemical Protective Glove Set.
Eye protection	Wear chemical goggles; use goggles and faceshield for splash hazards.
Other	Wear gloves and lab coat with M9 or M14 mask readily available for general lab work.
Emergency procedures	***Inhalation:*** Hold breath and don respiratory protection mask; if severe signs of agent exposure appear, administer immediately, in rapid succession, all three Nerve Agent Antidote Kits, Mark I injectors; use mouth-to-mouth resuscitation when approved mask-bag or oxygen delivery systems are not available; do not use mouth-to-mouth resuscitation when facial contamination exists; administer oxygen if breathing is difficult; seek medical attention immediately.
	Eye Contact: Flush eyes immediately with water for 10–15 min then don a respiratory protective mask. Although miosis may be an early sign of agent exposure, do not administer an injection when miosis is the only sign present; seek medical attention immediately.
	Skin Contact: Don respiratory mask and remove contaminated clothing; wash contaminated skin with copious amounts of soap and water immediately using 10% sodium carbonate solution, or 5% liquid household bleach; rinse well with water to remove decontamination; if local sweating and muscular symptoms occur, administer an intramuscular injection with the MARK I Kit; seek medical attention immediately.
	Ingestion: Do not induce vomiting; first symptoms are likely to be gastrointestinal; administer immediately 2-mg

intramuscular injection of the MARK I kit auto injectors; seek medical attention immediately.

Military Significant Information

Field protection Protective mask and protective clothing. Clothing gives off G-agents for about 30 min after contact with vapor; consider this fact before unmasking. Immediately remove all liquid from clothing.

Decontamination Flush eyes with water immediately. Use, the M258A1, M258, or M291 skin decontamination kit for liquid agent on the skin. Decontaminate individual equipment with the M280 individual equipment decontamination kit. HTH, STB, household bleach, caustic soda, dilute alkali solutions, or decontaminating solution number 2 (DS2) are effective on equipment. Use steam and ammonia or hot, soapy water in a confined area. Note: GA may react to form CK in bleach slurry.

Persistency The persistency will depend upon munitions used and the weather. Heavily splashed liquid persists 1–2 days under average weather condition. GA evaporates about 20× more slowly than water. GA in water can persist about 1 day at 20°C and about 6 days at 5°C. GA persists about twice as long in seawater.

Use Quick-acting casualty agent.

Chapter 6

Tear Agents

The tear compounds (lachrymators) cause a flow of tears and irritation of the skin. Because tear compounds produce only transient casualties, they are widely used for training, riot control, and situations where long-term incapacitation is unacceptable. When used against poorly equipped guerrilla or revolutionary armies, these compounds have proved extremely effective. When released indoors, they can cause serious illness or death.

The standard tear-producing agents currently in the US Army inventory for riot control are CS, CS1, CS2, CSX, and CR. The United States considers agent CN (popularly known as mace or tear gas) and its mixtures with various chemicals obsolete for military employment. This chapter includes these materials, however, for complete coverage of compounds with potential for use against US forces. This chapter also presents information regarding CN mixtures as an example of how agent properties can be tailored to the method of dissemination.

Riot control agents are irritants characterized by:

- A rapid time of onset of effects (seconds to several minutes).
- A relatively brief duration of effects (15–30 min) once the victim has escaped the contaminated atmosphere and has decontaminated (i.e., removed the material from his clothing).
- A high safety ratio (the ratio of the lethal dose [estimated] to the effective dose).

Orthochlorobenzylidene malononitrile (CS) is the most commonly used irritant for riot control purposes. Chloracetophenone (CN) is also used in some countries for this purpose in spite of its higher toxicity. A newer agent is dibenzoxazepine (CR) with which there is little experience. Arsenical smokes (sternutators) have in the past been used on the battlefield. Apart from their lachrymatory action they also provoke other effects, e.g., bronchoconstriction and emesis, and are some times referred to as vomiting agents.

6.1 Bromobenzylcyanide – CA

CA was the first tear agent that came into existence at the end of World War I. It was outmoded in 1920 with the introduction of the CN series and is now obsolete. The tear compounds cause a flow of tears and irritation of the skin. Because tear compounds produce only transient casualties, they are widely used for training, riot control, and situations where long-term incapacitation is unacceptable. When used against poorly equipped guerrilla or revolutionary armies, these compounds have proved extremely effective. When released indoors, they can cause serious illness or death.

Informational

Designation	CA
Class	Tear agent
Type	A – nonpersistent
Chemical name	Bromobenzylcyanide
CAS number	[5798-79-8]

Chemical and Physical Properties

Appearance	Pure bromobenzylcyanide is a colorless crystalline solid
Odor	A sour or rotten fruit odor
Chemical formula	C_8H_6BrN
Molecular weight	196.0
Chemical structure	
Melting point	25.5°C pure; 18.5°C plant purity
Boiling point	242°C (decomposes)
Flash point	None. Decomposes but does not burn
Decomposition temperature	Decomposes slowly at 60°C, more rapidly as the temperature increases. Decomposes completely at 242°C. Forms hydrobromic acid and dicyanostilbene.
Vapor density	6.7 (air = 1)
Liquid density	1.47 g/cm³ at 25°C
Solid density	1.52 g/cm³ at 20°C
Vapor pressure	0.011 mmHg at 20°C
Volatility	17 mg/m³ at 0°C

115 mg/m^3 at 20°C
217 mg/m^3 at 30°C

Solubility   Soluble in organic liquids. Insoluble in water and cold alcohol; also, soluble in phosgene, chloropicrin, and benzyl cyanide.

Reactivity

Hydrolysis products   Complex condensation products

Rate of hydrolysis   Very slow

Stability   No data available

Storage stability   Stable in glass, lead-lined, or enamel-lined containers; reaction with iron may be explosive. Vigorous corrosive action on all common metals except lead; reaction with iron may be explosive.

Decomposition   No data available

Polymerization   No data available

Toxicity

LD$_{50}$ (skin)   No data available

LCt$_{50}$ (respiratory)   8,000–11,000 mg-min/m^3 (estimated)

LCt$_{50}$ (percutaneous)   No data available

ICt$_{50}$ (respiratory)   30 mg-min/m^3 (approximate)

ICt$_{50}$ (percutaneous)   No data available

Rate of detoxification   Rapidly detoxifies at the low concentration ordinarily encountered.

Skin and eye toxicity   Irritating; not toxic

Rate of action   Instantaneous

Overexposure effects   CA is usually used in solution with ether or acetone as an aerosol. It produces a severe burning sensation to the mucous membranes and equally severe lacrimination to the eyes accompanied by headache and nausea. The nausea may lead to vomiting although the vomiting is more of a psychological reaction than physiological. CA will go into solution with human sweat and will incur a burning sensation to the face, especially in the areas around the mouth, nose, and eyes. It will penetrate clothing, making the areas around the neck, armpits, the tender skin areas behind the elbows, knees, and around the buttocks and crotch susceptible to rashes and blisters. The vapors can be lethal in enclosed or confined spaces within a few minutes without

5.2. Nerve Agent – A-232

This series of agents are a new class of nerve agents developed by the former Soviet Union. Very little information is available about them. It has been reported that this class of agents are 5–8, possibly as much as 10× stronger as VX. They are conjectured to be the unitary nerve agents from the binary novichoks. A-232 maybe the result of novichok-# and novichok-5 combining.

Informational

Designation	A-232
Class	Nerve agent
Type	No data available
Chemical name	[[(2-chloro-1-methylethoxy)fluorohydroxy phosphinyl]oxy]carbonimidic chloride fluoride
CAS number	[26102-98-7]

Chemical and Physical Properties

Appearance	No data available
Odor	No data available
Chemical formula	$C_4H_6Cl_2F_2NO_3P$
Molecular weight	255.97

Chemical structure

$$Cl-CH_2-CH(CH_3)-O-P(F)(=O)-O-N=C(Cl)-F$$

Melting point	No data available
Boiling point	No data available
Flash point	No data available
Decomposition temperature	No data available
Vapor density	No data available
Liquid density	No data available
Solid density	No data available
Vapor pressure	No data available
Volatility	No data available
Solubility	No data available

prior respiratory protection. Nausea can lead to unconsciousness which may mean suffocation.

Safety

Protective gloves	Wear butyl toxicological agent protective gloves (M3, M4, or glove set).
Eye protection	Wear chemical goggles; wear a mask/respirator in open areas.
Other	Wear a complete set of protective clothing to include gloves and lab coat with a respiratory mask readily available.
Emergency procedures	**Inhalation:** Remove victim from the source immediately; seek medical attention immediately.
	Eye Contact: Don a respiratory protective mask; flush eyes immediately with copious amounts of water; seek medical attention immediately.
	Skin Contact: Remove victim from the source immediately; decontaminate the skin immediately with copious amounts of water; decontaminate clothing with steam or by boiling; 20% alcoholic caustic soda is effective on material, but may damage it; seek medical attention immediately.
	Ingestion: Give victim milk to drink; seek medical attention immediately.

Military Significant Information

Field protection	Protective mask.
Decontamination	Decontaminate clothing with steam or by boiling. Twenty percent alcoholic caustic soda is effective on materials, but may damage it. Porous surfaces, such as earth, are very difficult to decontaminate.
Persistency	Depends on munitions used and the weather; heavily splashed liquid persists 1–2 days under average weather conditions.
Use	Obsolete.

6.2. Chloroacetophenone – CN

The United States considers agent CN (popularly known as mace or tear gas) and its mixtures with various chemicals to be obsolete for military deployment. It is highly toxic by inhalation and ingestion. CN tear compound causes flow of tears and irritation of the skin. Since tear compounds produce only transient casualties, they are wisely used for training, riot control, and situations where long-term incapacitation is unacceptable.

Informational

Designation	CN
Class	Tear agent
Type	A – nonpersistent
Chemical name	Chloroacetophenone
CAS number	[532-27-4]

Chemical and Physical Properties

Appearance	CN is a colorless to gray crystalline solid
Odor	A sharp, irritating floral odor
Chemical formula	C_8H_7ClO
Molecular weight	154.59

Chemical structure

Melting point	54°C pure; 46–48°C plant purity
Boiling point	248°C
Flash point	118°C
Decomposition temperature	Stable to boiling point
Vapor density	5.3 (air = 1)
Liquid density	1.187 g/cm³ at 58°C
Solid density	1.318 g/cm³ at 20°C
Vapor pressure	0.0026 mmHg at 0°C 0.0041 mmHg at 20°C 0.152 mmHg at 51.7°C
Volatility	2.36 mg/m³ at 0°C 34.3 mg/m³ at 20°C 1060 mg/m³ at 51.7°C
Solubility	Soluble in chloroform, chloropicrin, and other organic solvents; insoluble in water.

Reactivity

Hydrolysis products	Hydrochloric acid and hydroxyacetophenone
Rate of hydrolysis	Not readily hydrolyzed
Stability	No data available

Storage stability	Stable in closed containers at room temperature under normal storage and handling conditions. Incompatibility with water or steam.
Decomposition	Toxic and corrosive vapors are produced when combined with water or steam.
Polymerization	No data available

Toxicity

LD$_{50}$ (skin)	No data available
LCt$_{50}$ (respiratory)	7,000–14,000 mg-min/m^3
LCt$_{50}$ (percutaneous)	No data available
ICt$_{50}$ (respiratory)	80 mg-min/m^3
ICt$_{50}$ (percutaneous)	No data available
Rate of detoxification	Rapid; effects disappear in minutes. High concentrations may cause skin irritation that usually disappears within a few hours.
Skin and eye toxicity	Irritating; not toxic in concentrations likely to be encountered in the field.
Rate of action	Practically instantaneous
Overexposure effects	Alpha-chloroacetophenone vapors may cause a tingling or runny nose, burning and/or pain of the eyes, blurred vision, and tears. Burning in the chest, difficult breathing, and nausea may also occur as well as skin irritation, rash, or burns. It can also cause difficulty if swallowed.

Safety

Protective gloves	Wear impervious gloves.
Eye protection	Wear dust- and splash-proof safety goggles where there is any possibility of solid CN or liquids containing CN may contact the eyes; wear face shield; wear appropriate protective mask.
Other	Wear a complete set of protective clothing to include gloves and lab coat, apron, boots, and plastic coveralls; other protective clothing and equipment should be available to prevent contact with skin or clothing; remove contaminated clothing immediately; do not wear clothing until it has been properly laundered.
Emergency procedures	***Inhalation:*** Remove the victim to fresh air immediately; perform artificial respiration if breathing has stopped; keep victim warm and at rest; seek medical attention immediately.
	Eye Contact: Wash eyes immediately with copious amounts of water, lifting the lower and upper lids

occasionally; do not wear contact lenses when working with this chemical; seek medical attention immediately.

Skin Contact: Wash the contaminated skin using soap or mild detergent and water immediately; remove the contaminated clothing immediately and wash the skin using soap or mild detergent and water; seek medical attention immediately when there are chemical burns or evidence of skin irritation.

Ingestion: Induce vomiting by having victim touch the back of the throat with finger or by giving victim syrup of ipecac as directed; do not induce vomiting if victim is unconscious; seek medical attention immediately.

Military Significant Information

Field protection	Protective mask
Decontamination	Aeration in the field; soda ash solution or alcoholic caustic soda in enclosed areas.
Persistency	Short because the compounds are disseminated as an aerosol.
Use	Training and control of civil disturbances.

6.3. Chloropicrin – PS

PS was used in large quantities during World War I; it was stockpiled during World War II and is no longer authorized for military use. PS is more toxic than chlorine, but less toxic than phosgene (CG). PS is a severe respiratory irritant. Persons with impaired pulmonary function may be at increased risk from exposure. It is a possible but unconfirmed tumorigenic agent that decomposes to form toxic chlorine gas and nitrogen oxides near oxygen fires.

Informational

Designation	PS
Class	Tear agent
Type	A – nonpersistent
Chemical name	Trichloronitromethane
CAS number	[76-06-2]

Chemical and Physical Properties

Appearance	PS is a colorless, oily liquid
Odor	A stinging pungent odor
Chemical formula	CCl_3NO_2

Molecular weight	164.3
Chemical structure	$$Cl-\overset{\displaystyle Cl}{\underset{\displaystyle Cl}{C}}-NO_2$$
Melting point	–69°C
Boiling point	112°C
Flash point	Does not flash
Decomposition temperature	>400°C
Vapor density	5.6 (air = 1)
Liquid density	1.66 g/cm^3 at 20°C
Solid density	No data available
Vapor pressure	20 mmHg at 20°C
Volatility	55,700 mg/m^3 at 0°C 99,000 mg/m^3 at 10°C 164,500 mg/m^3 at 20°C 210,700 mg/m^3 at 25°C 267,500 mg/m^3 at 30°C
Solubility	Insoluble in water; soluble in organic solvents, lipids, organophosphorus compounds, mustards, phosgene, diphosgene, and Cl$_2$.

Reactivity

Hydrolysis products	Does not hydrolyzed in water
Rate of hydrolysis	Does not hydrolyzed in water
Stability	Contact with strong oxidizers may cause fires or explosions.
Storage stability	Instability occurs with high temperatures or severe shock, particularly when involving containers of greater than 30 gal capacity; unstable liquid. Liquid chloropicrin will attack some forms of plastics, rubber, and coatings.
Decomposition	Toxic gases and vapors (such as oxides of nitrogen, phosgene, nitrosyl chloride, chlorine, and carbon monoxide) may be released when chloropicrin decomposes.
Polymerization	No data available

Toxicity

LD$_{50}$ (skin)	No data available
LCt$_{50}$ (respiratory)	2,000 mg-min/m^3

LCt$_{50}$ (percutaneous)	No data available
ICt$_{50}$ (respiratory)	No data available
ICt$_{50}$ (percutaneous)	No data available
Rate of detoxification	No data available
Skin and eye toxicity	Irritates nose and throat; causes tearing; irritates lungs at higher concentrations; causes nausea and vomiting; can cause skin lesions.
Rate of action	No data available
Overexposure effects	Chloropicrin is a powerful irritant whose vapors cause lung, skin, eye, nose and throat irritation, coughing, and vomiting. As an eye irritant, it produces immediate burning, pain, and tearing. In high concentration, PS damages the lungs, causing pulmonary edema. Exposure to liquid PS can cause severe burns on the skin that generally result in blisters and lesions. The lowest irritant concentration is 9 mg-min/m^3 for 10 min, and the median lethal concentration is 2,000 mg-min/m^3.

Safety

Protective gloves	Wear impervious gloves.
Eye protection	Wear face shields (8 in. minimum) or dust- and splash-proof safety goggles to prevent any possibility of skin contact with liquid chloropicrin.
Other	Wear a complete set of protective clothing to include gloves and lab coat, apron, boots, plastic coveralls; other protective clothing and equipment should be available to prevent contact with the skin or clothing; remove contaminated clothing immediately, do not wear clothing until it has been properly laundered.
Emergency procedures	**Inhalation:** Remove the victim to fresh air immediately; perform artificial respiration if breathing has stopped; keep the victim warm and at rest; seek medical attention immediately.
	Eye Contact: Wash eyes immediately with copious amounts of water, lifting the lower and upper lids occasionally; do not wear contact lenses when working with this chemical; seek medical attention immediately.
	Skin Contact: Wash the contaminated skin using soap or mild detergent and water; remove the contaminated clothing immediately; wash the skin using soap or mild detergent and water; if irritation persists after washing, seek medical attention immediately.
	Ingestion: Give victim copious amounts of water immediately; induce vomiting by having victim touch the back of his throat with his finger; do not make an unconscious person vomit; seek medical attention immediately.

Military Significant Information

Field protection	Protective mask for vapors.
Decontamination	Neutral or slight basic solutions with sulfides, such as sodium sulfide. Do not use acidic solutions for decontamination; acids reduce PS to CX, a blister agent.
Persistency	Approximately 6 h in vegetated fields.
Use	Not authorized for US military use.

6.4. o-Chlorobenzylidene Malononitrile – CS

CS was developed in the late 1950s as a riot control substance. It is a more potent irritant than chloroacetophenone, but is less incapacitating. In the late 1960s, stocks of CS replaced CN. Presently, the US Army uses CS for combat training and riot control purposes.

Informational

Designation	CS
Class	Tear agent
Type	B – persistent
Chemical name	o-Chlorobenzylidene malononitrile
CAS number	[2698-41-1]

Chemical and Physical Properties

Appearance	CS is a white crystalline solid
Odor	Burnt to create a colorless gas with an acrid pepper-like smell.
Chemical formula	$C_{10}H_5ClN_2$
Molecular weight	188.61
Chemical structure	

Melting point	93–95°C
Boiling point	310–315°C
Flash point	197°C
Decomposition temperature	Unknown

Vapor density	6.5 (air = 1)
Liquid density	No data available
Solid density	1.04 g/cm^3 at 20°C
Vapor pressure	3.4×10^{-5} mmHg at 20°C
Volatility	0.71 mg/m^3 at 25°C
Solubility	Soluble in hexane, benzene, methylene chloride, acetone, dioxane, ethyl acetate, and pyridine; insoluble in water and ethanol.

Reactivity

Hydrolysis products	o-chlorobenzaldehyde and malonoitrile
Rate of hydrolysis	Rapid for dissolved CS. CS is only slightly soluble in water (about 0.008% by wt at 25°C); thus, solid CS in water hydrolyzes relatively slowly. CS hydrolyzes more rapidly in alkalinity is increased.
Stability	Incompatible with strong oxidizers
Storage stability	Stable in storage
Decomposition	When heated to decomposition, CS emits highly toxic fumes.
Polymerization	No data available

Toxicity

LD_{50} *(skin)*	No data available
LCt_{50} *(respiratory)*	61,000 $mg\text{-}min/m^3$
LCt_{50} *(percutaneous)*	No data available
ICt_{50} *(respiratory)*	10–20 $mg\text{-}min/m^3$
ICt_{50} *(percutaneous)*	No data available
Rate of detoxification	Quite rapid. Incapacitating dosages lose their effects in 5–10 min.
Skin and eye toxicity	Highly irritating, but not toxic.
Rate of action	Very rapid (maximum effects in 20–60 s).
Overexposure effects	CS is disseminated by burning, explosion, and aerosol formation. It is immediately irritating to the eyes and upper respiratory tract. Warm vapors mix with human sweat to cause a burning sensation to the eyes, nose, and mouth. Conjunctivitis and pain in the eyes, lacrimation, erythema of the eyelids, runny nose, burning throat, coughing, and constricted feeling in the chest are the effects which will

occur immediately and will persist 5–20 min after removal from the contaminated area. It is immediately dangerous to life and health at a concentration of 2 mg/m^3. It is not an accumulative agent in the human body, although it accumulates in the landscape. CS is the most persistent of the tear agents, absorbing into the most porous surfaces including soil and plaster.

Safety

Protective gloves	Wear impervious gloves; rubber gloves.
Eye protection	Wear face shields or dust- and splash-proof safety goggles to prevent any possibility of skin contact.
Other	Wear protective mask and overclothing in confined spaces; use a chemical cartridge respirator with organic vapor cartridges in combination with a high efficiency particulate filter; wear a self-contained breathing apparatus with a full-face piece or an air purifying, full-face piece respirator with an organic vapor canister.
Emergency procedures	**Inhalation:** Remove the victim to fresh air immediately; perform artificial respiration if breathing has stopped; keep the victim warm and at rest; seek medical attention immediately.
	Eye Contact: Wash eyes immediately with copious amounts of water for at least 15 min; apply an ophthalmic corticosteroid ointment after decontamination; treat delayed erythema with a bland shake lotion (such as calamine lotion) or a topical corticosteroid depending on severity; do not wear contact lenses when working with this chemical; seek medical attention immediately.
	Skin Contact: Wash the contaminated skin thoroughly using soap and water; remove the contaminated clothing immediately; if irritation persists after washing, seek medical attention immediately.
	Ingestion: Give victim copious amounts of water immediately; induce vomiting by having victim touch the back of throat with finger; do not make an unconscious person vomit; seek medical attention immediately.

Military Significant Information

Field protection	Protective mask and ordinary field clothing secured at neck, wrist, and ankles. Personnel handling CS should wear rubber gloves for additional protection.
Decontamination	Personnel affected by CS in field concentrations should move to an uncontaminated area, face into the wind, and remain well-spaced. They should be warned not to rub their

eyes to scratch irritated skin areas. Normally, aeration is sufficient to decontaminate personnel and to dissipate ill effects of the compound in 5–10 min. Personnel contaminated with visible CS particles should flush their bodies or affected parts with cool water for 3–5 min before showering with warm or hot water. Use soap and water on equipment contaminated with CS. The higher the alkalinity of the soap, the better. Do not use standard decontaminants or detergents that contain chlorine bleach for CS containing compounds; the materials can react to form compounds (epoxides, which have vesicant properties) more toxic than CS.

Persistency   Varies, depending upon amount of contamination and form of CS. CS aerosol usually has little residual hazard.

Use   Training and riot control; limited tactical use in counter-guerrilla operations.

6.5. Tear Agent – CNB

The symbol CNB identifies a mixture of 10% CN, 45% carbon tetrachloride, and 45% benzene. It is a powerful lacrimator. US forces adopted CNB in 1920 and used it until CNS replaced it. The advantage claimed for CNB was that its lower CN content made it more satisfactory than CNC for training purposes. Actually, merely using a lower concentration would obtain the same result with CNC.

Informational

Designation   CNB

Class   Tear agent

Type   A – nonpersistent

Chemical name   Chloroacetophenone; carbon tetrachloride; and benzene

CAS number   Mixture: [not available]; chloroacetophenone: [532-27-4]; carbon tetrachloride: [56-23-5]; benzene: [71-43-2].

Chemical and Physical Properties

Appearance   CNB is a slightly brown liquid

Odor   Smells heavily of benzene

Chemical formula C_8H_7ClO; CCl_4; C_6H_6

Molecular weight   154.59; 154; 78

Chemical structure

$$Cl-\underset{\underset{Cl}{|}}{\overset{\overset{Cl}{|}}{C}}-Cl$$

Melting point	–7 to 30°C
Boiling point	75–247°C, varies as solvents evaporate
Flash point	Below 4.44°C
Decomposition temperature	>247°C
Vapor density	4.0 (air = 1)
Liquid density	1.14 g/cm^3 at 20°C
Solid density	No data available
Vapor pressure	No data available
Volatility	Benzene: 320,624 mg/m^3 at 20°C 420,111 mg/m^3 at 25°C Carbon tetrachloride: 766,000 mg/m^3 at 20°C
Solubility	No data available

Reactivity

Hydrolysis products	None
Rate of hydrolysis	None
Stability	Stable
Storage stability	Adequate in storage
Decomposition	No data available
Polymerization	No data available

Toxicity

LD_{50} *(skin)*	No data available
LCt_{50} *(respiratory)*	11,000 mg-min/m^3
LCt_{50} *(percutaneous)*	No data available
ICt_{50} *(respiratory)*	80 mg-min/m^3
ICt_{50} *(percutaneous)*	No data available
Rate of detoxification	Rapid
Skin and eye toxicity	Irritating, not toxic
Rate of action	Instantaneous

Overexposure effects   CNB is a formulation of chloroacetophenone. Like CN, CNB has a pronounced lacrimatory effect, resulting in a natural reflex to shut the eyes. It is similarly irritating to the skin, especially the face and such exposed portions of the body where sweat accumulates. CNB will penetrate clothing or adhere to it under long exposure due to its benzene component. The same rashes caused by CN will be caused by CNB. CNB has a slightly more powerful choking effect than CN. Eye toxicity remains about the same as CN. Some sensitive individuals may experience nausea upon exposure. CNB can form lethal concentrations in closed or confined spaces, although concentrations in open are highly unlikely ever to do so.

Safety

Protective gloves   Wear impervious gloves.

Eye protection   Wear dust- and splash-proof safety goggles where there is any possibility of solid CNB or liquids containing CNB contacting the eyes; use appropriate protective mask.

Other   Wear a complete set of protective clothing to include gloves and lab coat, apron, boots, plastic coveralls; other protective clothing and equipment should be available to prevent contact with skin or clothing; remove contaminated clothing immediately; do not wear clothing until it has been properly laundered.

Emergency procedures   **Inhalation:** Remove the victim to fresh air immediately; perform artificial respiration if breathing has stopped; keep victim warm and at rest; seek medical attention immediately.

Eye Contact: Wash eyes immediately with copious amounts of water, lifting the lower and upper lids occasionally; do not wear contact lenses when working with this chemical; seek medical attention immediately.

Skin Contact: Wash the contaminated skin with soap or mild detergent and water immediately; remove the contaminated clothing immediately; wash the skin using soap or mild detergent and water; seek medical attention immediately when there are chemical burns or evidence of skin irritation.

Ingestion: Induce vomiting by having victim touch the back of his throat with finger or by giving victim syrup of ipecac as directed; do not induce vomiting if victim is unconscious; seek medical attention immediately.

Military Significant Information

Field protection   Protective mask for vapors

Decontamination   None needed in the field; wash contaminated surfaces with a 5% solution by weight of sodium hydroxide in 95%

alcohol or with a mixture of 20 parts of water and 80 parts of carbitol (diethylene glycol).

Persistency	Short
Use	Obsolete

6.6. Tear Agent – CNC

The symbol CNC identifies a 30% solution of CN in chloroform. It was developed to deliver CN in liquid form. CNC causes a flow of tears, irritates the respiratory system, and causes stinging of skin.

Informational

Designation	CNC
Class	Tear agent
Type	A – nonpersistent
Chemical name	Chloroacetophenone; chloroform
CAS number	Mixture: [not available]; chloroacetophenone: [532-27-4]; chloroform: [67-66-4]

Chemical and Physical Properties

Appearance	CNC is a liquid
Odor	Smells heavily of chloroform
Chemical formula	C_8H_7ClO; $CHCl_3$
Molecular weight	154.59; 119.4
Chemical structure	
Melting point	0.23°C. This is the temperature at which CN crystals separate and is not a true change of state. It is that temperature at which the solution becomes saturated with CN. If the solution cools below this point, solid matter appears and gives the appearance of solidifying.
Boiling point	60–247°C, varies as solvents evaporate
Flash point	None
Decomposition temperature	Stable to boiling point

Vapor density	4.4 (air = 1)
Liquid density	1.40 g/cm^3 at 20°C
Solid density	No data available
Vapor pressure	61 mmHg at 5°C
	144 mmHg at 20°C
Volatility	No data available
Solubility	No data available

Reactivity

Hydrolysis products	Hydrogen chloride and hydroxyacetophenone
Rate of hydrolysis	Not readily hydrolyzed
Stability	Stable
Storage stability	Adequate in storage
Decomposition	No data available
Polymerization	No data available

Toxicity

LD$_{50}$ (skin)	No data available
LCt$_{50}$ (respiratory)	11,000 mg-min/m^3
LCt$_{50}$ (percutaneous)	No data available
ICt$_{50}$ (respiratory)	80 mg-min/m^3
ICt$_{50}$ (percutaneous)	No data available
Rate of detoxification	Rapid
Skin and eye toxicity	Irritating, not toxic
Rate of action	Instantaneous
Overexposure effects	CNC is a formulation of chloroacetophenone. Like CN, CNC has a pronounced lacrimatory effect, resulting in a natural reflex to shut the eyes. It is similarly irritating to the skin, especially the face and such exposed portions of the body where sweat accumulates. The same rashes caused by CN will be caused by CNC. CNC has a slightly more powerful choking effect than CN. Eye toxicity remains about the same as CN. Some sensitive individuals may experience nausea upon exposure. CNC can form lethal concentrations in closed or confined spaces, although concentrations in open are highly unlikely ever to do so.

Safety

Protective gloves	Wear impervious gloves.
Eye protection	Wear dust- and splash-proof safety goggles where there is any possibility of solid CNC or liquids containing CNC contacting the eyes; use appropriate protective mask.
Other	Wear a complete set of protective clothing to include gloves and lab coat, apron, boots, plastic coveralls; other protective clothing and equipment should be available to prevent contact with skin or clothing; remove contaminated clothing immediately; do not wear clothing until it has been properly laundered.
Emergency procedures	***Inhalation:*** Remove the victim to fresh air immediately; perform artificial respiration if breathing has stopped; keep victim warm and at rest; seek medical attention immediately.
	Eye Contact: Wash eyes immediately with copious amounts of water, lifting the lower and upper lids occasionally; do not wear contact lenses when working with this chemical; seek medical attention immediately.
	Skin Contact: Wash the contaminated skin with soap or mild detergent and water immediately; remove the contaminated clothing immediately; wash the skin using soap or mild detergent and water; seek medical attention immediately when there are chemical burns or evidence of skin irritation.
	Ingestion: Induce vomiting by having victim touch the back of his throat with finger or by giving victim syrup of ipecac as directed; do not induce vomiting if victim is unconscious; seek medical attention immediately.

Military Significant Information

Field protection	Protective mask for vapors
Decontamination	Aeration in the field; wash contaminated surfaces with a 5% solution by weight of sodium hydroxide in 95% alcohol or strong soda ash solution.
Persistency	Short
Use	Obsolete

6.7. Tear Agent – CNS

The symbol CNS identifies a mixture of 23% CN, 38.4% chloropicrin, and 38.4% chloroform. It is an example of multiple-component mixtures developed to achieve desired dissemination characteristics. CNS was declared obsolete in 1957 and is no longer in the supply system. In addition to having the effects described under CN, CNS also had the effects of PS, which acts as a vomiting compound, a choking agent, and a tear compound. CNS may cause lung effects similar to those of phosgene and also may cause nausea, vomiting, colic, and diarrhea that may persist for weeks. The lacrimatory effects of PS are much less marked than those of CN and were relatively unimportant for CNS. This is shown by the fact that tearing effects were no greater with CNS than with CNC, which contains no PS.

Informational

Designation	CNS
Class	Tear agent
Type	A – nonpersistent
Chemical name	Chloroacetophenone; chloropicrin; chloroform
CAS number	Mixture: [not available]; chloroacetophenone: [532-27-4]; chloropicrin: [76-06-2]; chloroform: [67-66-3].

Chemical and Physical Properties

Appearance	CNS is a clear liquid
Odor	Smells like flypaper
Chemical formula	C_8H_7ClO; CCl_3NO_2; $CHCl_3$
Molecular weight	154.59; 164.3; 119.4

Chemical structure

Melting point	Approximately 2°C This is the temperature at which CN crystals separate and is not a true change of state.
Boiling point	60–247°C, varies as solvents evaporate.
Flash point	None
Decomposition temperature	Stable to boiling point
Vapor density	5.0 (air = 1)
Liquid density	1.47 g/cm³ at 20°C
Solid density	No data available

Vapor pressure	78 mmHg at 5°C
Volatility	605,000 mg/m^3 at 20°C
	900,500 mg/m^3 at 30°C
	1,620,000 mg/m^3 at 50°C
Solubility	No data available

Reactivity

Hydrolysis products	Hydrogen chloride and hydroxyacetophenone
Rate of hydrolysis	Not readily hydrolyzed
Stability	Stable
Storage Stability	Adequate in storage
Decomposition	No data available
Polymerization	No data available

Toxicity

LD$_{50}$ (skin)	No data available
LCt$_{50}$ (respiratory)	11,400 mg-min/m^3
LCt$_{50}$ (percutaneous)	No data available
ICt$_{50}$ (respiratory)	60 mg-min/m^3
ICt$_{50}$ (percutaneous)	No data available
Rate of detoxification	The effects of CN are long lasting and cumulative and may prolong the effects of CNS for weeks. Such a prolonged effect maybe undesirable for training and riot control.
Skin and eye toxicity	Irritating, not toxic
Rate of action	Instantaneous
Overexposure effects	CNS is an example of multiple-component mixtures developed to achieve desired dissemination characteristics. Its hazards exist for inhalation, ingestion, and skin and eye exposure. It produces nausea within a minute of inhalation by a moderately sensitive person. If inhaled for longer periods, vomiting, colic (severe abdominal pains and cramps), and diarrhea are to be expected in its victims. Persons who are exposed to very large quantities of the vapors or liquid concentrations may suffer these symptoms for weeks. CNS is a nonlethal choking agent. It will cause the victims to gasp for air (thus inhaling more CNS) while causing discomfort to the bronchial tubes and lung sacs. CNS vapors may go into solution with sweat, making it a skin irritant, especially

the face. If allowed to penetrate the clothing, CNS will cause stinging under the armpits, elbows, knees, and the area around the crotch and buttocks. Skin rashes may result after prolonged exposures. Prolonged eye exposure would not be recommended.

Safety

Protective gloves	Wear impervious gloves.
Eye protection	Wear dust- and splash-proof safety goggles where there is any possibility of solid CNS or liquids containing CNS contacting the eyes; use appropriate protective mask.
Other	Wear a complete set of protective clothing to include gloves and lab coat, apron, boots, plastic coveralls; other protective clothing and equipment should be available to prevent contact with skin or clothing; remove contaminated clothing immediately; do not wear clothing until it has been properly laundered.
Emergency procedures	***Inhalation:*** Remove the victim to fresh air immediately; perform artificial respiration if breathing has stopped; keep victim warm and at rest; seek medical attention immediately.
	Eye Contact: Wash eyes immediately with copious amounts of water, lifting the lower and upper lids occasionally; do not wear contact lenses when working with this chemical; seek medical attention immediately.
	Skin Contact: Wash the contaminated skin with soap or mild detergent and water immediately; remove the contaminated clothing immediately, and wash the skin using soap or mild detergent and water; seek medical attention immediately when there are chemical burns or evidence of skin irritation.
	Ingestion: Induce vomiting by having victim touch the back of his throat with finger or by giving victim syrup of ipecac as directed; do not induce vomiting if victim is unconscious; seek medical attention immediately.

Military Significant Information

Field protection	Protective mask for vapors
Decontamination	Not required in the field, hot solution of soda ash and sodium sulfite for gross contamination in enclosed spaces.
Persistency	Short
Use	Obsolete

6.8. Tear Agent – CR

In 1974 the US Army approved the use of CR. CR has much greater irritating properties than CS and is about 5× more effective. In addition, CR is much less toxic than CS. CR is not used in its pure form (a yellow powder), but is dissolved in a solution of 80 parts of propylene glycol and 20 parts of water to form a 0.1% CR solution. It is used in solution as a riot control agent.

Informational

Designation	CR
Class	Tear agent
Type	B – persistent
Chemical name	Dibenz-(b,f)-1,4-oxazepine
CAS number	[257-07-8]

Chemical and Physical Properties

Appearance	Yellow powder
Odor	Burning sensation in nose and sinuses
Chemical formula	$C_{13}H_9NO$
Molecular weight	195.25
Chemical structure	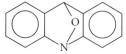
Melting point	72°C
Boiling point	335°C
Flash point	188°C
Decomposition temperature	No data available
Vapor density	6.7 (air = 1)
Liquid density	1.56 g/cm³ at 25°C
Solid density	No data available
Vapor pressure	0.00059 mmHg at 20°C
Volatility	0.63 mg/m³ at 25°C
Solubility	Negligible in water. Soluble in polypropylene glycol, benzene, chloroform, or carbon tetrachloride.

Reactivity

Hydrolysis products	No data available
Rate of hydrolysis	Very slow
Stability	Stable in aqueous, heated acidic, and strong alkali solutions. Thermally stable. Never mix CR with organic or inorganic bleaches or peroxide.
Storage stability	Store in nonporous containers on nonporous surfaces. CR will leach and persist in porous materials.
Decomposition	Nitrous oxide and carbon monoxide
Polymerization	No data available

Toxicity

LD_{50} *(skin)*	No data available
LCt_{50} *(respiratory)*	No data available
LCt_{50} *(percutaneous)*	No data available
ICt_{50} *(respiratory)*	No data available
ICt_{50} *(percutaneous)*	0.15 mg/m^3
Rate of detoxification	Skin irritation persists for 15–30 min after removal of CR. Eye effects may last up to 6 h.
Skin and eye toxicity	No data available
Rate of action	Very rapid
Overexposure effects	No data available

Safety

Protective gloves	Wear impervious gloves.
Eye protection	Wear dust- and splash-proof safety goggles where there is any possibility of solid CN or liquids containing CN may contact the eyes; wear face shield; wear appropriate protective mask.
Other	Wear a complete set of protective clothing that include gloves and lab coat, apron, boots, and plastic coveralls; other protective clothing and equipment should be available to prevent contact with skin or clothing; remove contaminated clothing immediately; do not wear clothing until it has been properly laundered.
Emergency procedures	**Inhalation:** Remove the victim to fresh air immediately; perform artificial respiration if breathing has stopped; keep victim warm and at rest; seek medical attention immediately.

Eye Contact: Wash eyes immediately with copious amounts of water, lifting the lower and upper lids occasionally; do not wear contact lenses when working with this chemical; seek medical attention immediately.

Skin Contact: Wash the contaminated skin using soap or mild detergent and water immediately; remove the contaminated clothing immediately and wash the skin using soap or mild detergent and water; seek medical attention immediately when there are chemical burns or evidence of skin irritation.

Ingestion: Induce vomiting by having victim touch the back of the throat with finger or by giving victim syrup of ipecac as directed; do not induce vomiting if victim is unconscious; seek medical attention immediately.

Military Significant Information

Field protection

Protective mask and ordinary field clothing secured at neck, wrist, and ankles.

Decontamination

Personnel affected by CR in field concentrations should move to an uncontaminated area, face into the wind, and remain well-spaced apart. They should be warned not to rub their eyes to scratch irritated skin areas. Normally, aeration is sufficient to decontaminate personnel and to dissipate ill effects of the compound in 5–10 min. If CR entered the eyes, flush with water. Individuals who have ingested CR should be given lots of water or milk to drink. Do not induce vomiting. Do not use bleach, detergents, or peroxides for decontamination; this combination releases toxic fumes. Remove CR from equipment or surfaces by wiping, scraping, shoveling, or sweeping. Wipe the area with rags soaked in propylene glycol or an automotive antifreeze solution, if available; wipe with rubbing alcohol; and then scrub with nonbleach detergent and hot water before rinsing with large amounts of cold water. Place all contaminated materials used for decontamination in a storage container where they cannot affect personnel.

Persistency

As persistent as, or more persistent than, CS. Under suitable conditions CR can persist on certain surfaces (especially porous material) for up to 60 days.

Use

Riot control agent dispersed as a spray.

Chapter 7

Vomit Agents

Vomiting agents produce strong, pepper-like irritation in the upper respiratory tract, with irritation of the eyes and tearing. They cause violent, uncontrollable sneezing, cough, nausea, vomiting, and a general feeling of bodily discomfort. The vomiting compounds listed are normally solids that vaporize when heated and then condense to form aerosols. They produce their effects by inhalation or by direct action on the eyes. Under field conditions, vomiting compounds cause great discomfort to victims; when released indoors, they can cause serious illness or death. The principal vomiting agents are diphenylchloroarsine (DA), diphenylaminochloroarsine (DM; Adamsite), and diphenylcyanoarsine (DC). Chloropicrin also is a vomiting agent.

The onset of symptoms may be delayed for several minutes after initial exposure (especially with DM); effective exposure may, therefore, occur before the presence of the smoke is suspected. The paranasal sinuses are irritated and fill with secretions and severe frontal headache results. Prolonged exposure may cause retrosternal pain, dyspnoea, and asthma-like symptoms. Symptoms reach their climax after 5–10 min and disappear 1–2 h after cessation of exposure.

Symptoms, in progressive order, are irritation of the eyes and mucous membranes, viscous discharge from the nose similar to that caused by a cold, sneezing and coughing, severe headache, acute pain and tightness in the chest, nausea, and vomiting. Exposure of the skin to high concentrations will cause erythema and itching, proceeding to a burning sensation and vesicle formation.

The effects of DM develop more slowly than those of DA, and for moderate concentrations the effects last about 30 min after the person leaves the contaminated atmosphere. At higher concentrations the effects may last up to several hours.

7.1. Adamsite – DM

DM was first produced during World War I. Adamsite was not toxic enough for the battlefield, but it proved to be too drastic for use against civilian mobs;

it was banned for use against civilian populations in the 1930s, Western nations. DM was produced worldwide until superseded by the CN series of tear agents.

Informational

Designation	DM
Class	Vomiting agent
Type	A – nonpersistent
Chemical name	10-chloro-5,10-dihydrophenarsazine chloride
CAS number	[578-94-9]

Chemical and Physical Properties

Appearance	Light green to yellow crystals at room temperature
Odor	No odor, but irritating
Chemical formula	$C_{12}H_9AsClN$
Molecular weight	277.57

Chemical structure

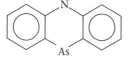

Melting point	195°C
Boiling point	410°C
Flash point	Does not flash
Decomposition temperature	Above melting point
Vapor density	Negligible
Liquid density	No data available
Solid density	1.65 g/cm³ at 20°C
Vapor pressure	2×10^{-13} mmHg at 20°C
Volatility	19,300 mg/m³ at 0°C
	26,000–120,000 mg/m³ at 20°C
	72,500–143,000 mg/m³ at 25°C
Solubility	Soluble in furfural and acetone; slightly soluble in common organic solvents; insoluble in water. Not readily soluble in any of the liquid chemical warfare agents.

Reactivity

Hydrolysis products   Diphenylarsenious oxide $[NH(C_6H_4)_2 As]_2O$ and hydrochloric acid. The oxide is very poisonous if taken internally.

Rate of hydrolysis   Quite rapid when in aerosol form. When solid DM is covered with water, a protective oxide coating forms that hinders further hydrolysis. Acidic pH: 0.5% prevents hydrolysis at room temperature; 9.8% HCl prevents hydrolysis at 70°C. Basic pH slowly hydrolyzes in water.

Stability   Stable

Storage stability   Stable in pure form; after 3 months, caused extensive corrosion of aluminum, anodized aluminum, and stainless steel; will corrode iron, bronze, and brass when moist. Titanium: 71°C, 6 months, appeared good. Stainless Steel: 43°C, 30 days, slight discoloration. Common Steel: 43°C, 30 days, covered with rust. Aluminum Anodized: 43°C, 30 days, minor corrosion and pitting. Aluminum: 43°C, 30 days, severe corrosion.

Decomposition   No data available

Polymerization   No data available

Toxicity

LD_{50} (skin)   No data available

LCt_{50} (respiratory)   11,000 mg-min/m^3

LCt_{50} (percutaneous)   No data available

ICt_{50} (respiratory)   22–150 mg-min/m^3

ICt_{50} (percutaneous)   No data available

Rate of detoxification   Quite rapid in small amounts. Incapacitating amounts lose their effects after about 30 min

Skin and eye toxicity   Irritating; relatively nontoxic

Rate of action   Very high. Requires only about 1 min to temporarily incapacitate at a concentration of 22 mg/m^3.

Overexposure effects   DM is a vomiting compound. It is normally a solid, but upon heating, DM first vaporizes and then condenses to form aerosols. It is toxic through inhalation, ingestion, and skin contact. Adamsite is dispersed as an aerosol, irritating to the eyes and respiratory tract, but not necessarily to the skin. Under field conditions, vomiting agents can cause great discomfort to the victims; when released indoors, they can cause serious illness or death. Symptoms include irritation of eyes and mucous membranes, coughing, sneezing, severe headache, acute pain and tightness in the chest, nausea, and vomiting. DM has been noted to cause necrosis of corneal epithelium in humans. The human body will detoxify the effects of mild exposures within 30 min of evacuation. Severe exposures may take several hours to detoxify and minor sensory disturbances may persist for up to 1 day.

Safety

Protective gloves	Wear chemical protective gloves.
Eye protection	Wear chemical goggles; wear a mask/respirator in open areas.
Other	Wear additional protective clothing, such as gloves and lab coat with an M9, M17, or M40 mask readily available in closed or confined spaces.
Emergency procedures	**Inhalation:** Remove victim to fresh air; wear a mask/respirator in spite of coughing, sneezing, salivation, and nausea; lift the mask from the face briefly, if necessary, to permit vomiting or to drain saliva from the facepiece; seek medical attention immediately.
	Eye Contact: Don a respiratory protective mask; seek medical attention immediately.
	Skin Contact: Rinse the nose and throat with saline water or bicarbonate of soda solution; wash exposed skin and scalp with soap and water and allow to dry on the skin; dust the skin with borated talcum.
	Ingestion: Seek medical attention immediately; carry on duties as vigorously as possible; this will help to lessen and shorten the symptoms; combat duties usually can be performed in spite of the effects of sternutators.

Military Significant Information

Field protection	Protective mask.
Decontamination	None needed in the field; use bleaching powder or DS2 for gross decontamination in enclosed places.
Persistency	Short, because compounds are disseminated as an aerosol. Soil; persistent. Surface (wood, metal, masonry, rubber, paint); persistent. Water; persistent; when material is covered with water, an insoluble film forms, which prevents further hydrolysis.
Use	Not authorized for US military use.

7.2. Diphenylchloroarsine – DA

Agent DA contains arsenic and chlorine. The threat could use it to cause troops to remove their masks and vomit, thereby exposing them to other agents.

Informational

Designation	DA
Class	Vomiting agent
Type	A – nonpersistent
Chemical name	Diphenylchloroarsine
CAS number	[712-48-1]

Chemical and Physical Properties

Appearance	Colorless crystals when pure
Odor	None
Chemical formula	$C_{12}H_{10}AsCl$
Molecular weight	264.5

Chemical structure

Melting point	41–44.5°C
Boiling point	333°C with decomposition
Flash point	350°C
Decomposition temperature	300°C
Vapor density	9.15 (air = 1)
Liquid density	1.387 g/cm³ at 50°C
Solid density	1.363 g/cm³ at 40°C
Vapor pressure	0.0036 mmHg at 45°C, calculated
Volatility	48 mg/m³ at 45°C
Solubility	Soluble in acetone, ethanol, and carbon tetrachloride; insoluble in water.

Reactivity

Hydrolysis products	Diphenylaresenious oxide and hydrochloric acid. The oxide is very poisonous if taken internally.
Rate of hydrolysis	Slow in mass, but rapid when finely divided.
Stability	No data available
Storage stability	Stable when pure.

| **Decomposition** | No data available |
| **Polymerization** | No data available |

Toxicity

LD_{50} **(skin)**	No data available
LCt_{50} **(respiratory)**	15,000 mg-min/m^3 (estimated)
LCt_{50} **(percutaneous)**	No data available
ICt_{50} **(respiratory)**	12 mg-min/m^3
ICt_{50} **(percutaneous)**	No data available
Rate of detoxification	Any merely incapacitating amount is detoxified completely within 1–2 h.
Skin and eye toxicity	Irritating; not toxic
Rate of action	Very rapid, within 2 or 3 min after a 1-min exposure.
Overexposure effects	No data available

Safety

Protective gloves	Wear chemical protective gloves.
Eye protection	Wear chemical goggles; wear a mask/respirator in open areas.
Other	Wear additional protective clothing, such as gloves and lab coat with an M9, M17, or M40 mask readily available in closed or confined spaces.
Emergency procedures	**Inhalation:** Remove victim to fresh air; wear a mask/respirator in spite of coughing, sneezing, salivation, and nausea; lift the mask from the face briefly, if necessary, to permit vomiting or to drain saliva from the facepiece; seek medical attention immediately.
	Eye Contact: Don a respiratory protective mask; seek medical attention immediately.
	Skin Contact: Rinse the nose and throat with saline water or bicarbonate of soda solution; wash exposed skin and scalp with soap and water and allow to dry on the skin; dust the skin with borated talcum.
	Ingestion: Seek medical attention immediately; carry on duties as vigorously as possible; this will help to lessen and shorten the symptoms; combat duties usually can be performed in spite of the effects of sternutators.

Military Significant Information

| **Field protection** | Protective mask. |
| **Decontamination** | None required in the field. Caustic soda or chlorine used for gross contamination in enclosed spaces. |

Persistency	Short, compound is disseminated as an aerosol.
Use	Not authorized for US military use.

7.3. Diphenylcyanoarsine – DC

The properties of DC are much like those of DA and DM, and the threat would use it in the same manner. DC is more toxic than DA. The effects from a moderate concentration last about 30 min after a person leaves the contaminated atmosphere. The effects from a higher concentration may last up to several hours.

Informational

Designation	DC
Class	Vomiting agent
Type	A – nonpersistent
Chemical name	Diphenylcyanoarsine
CAS number	[23525-22-6]

Chemical and Physical Properties

Appearance	White to pink solid
Odor	Similar to garlic and bitter almonds
Chemical formula	$C_{12}H_{10}AsN$
Molecular weight	255.0
Chemical structure	
Melting point	31.5–35°C
Boiling point	350°C with decomposition
Flash point	No data available
Decomposition temperature	About 25% decomposed at 300°C. Largely decomposed as a result of dispersing blast.
Vapor density	No data available
Liquid density	1.3338 g/cm^3 at 35°C
Solid density	No data available
Vapor pressure	0.0002 mmHg at 20°C
Volatility	2.8 mg/m^3 at 20°C
Solubility	Soluble in chloroform and other organic solvents; insoluble in water.

Reactivity

Hydrolysis products	Hydrochloric acid and diphenylarsenious oxide
Rate of hydrolysis	Very slow
Stability	Stable at all ordinary temperatures
Storage stability	Stable at all ordinary temperatures
Decomposition	No data available
Polymerization	No data available

Toxicity

LD_{50} *(skin)*	No data available
LCt_{50} *(respiratory)*	10,000 mg-min/m^3
LCt_{50} *(percutaneous)*	No data available
ICt_{50} *(respiratory)*	30 mg-min/m^3
ICt_{50} *(percutaneous)*	No data available
Rate of detoxification	Rapid. Incapacitating amounts lose their effect after about 1 h.
Skin and eye toxicity	Irritating; not toxic
Rate of action	Very rapid. Higher concentrations are intolerable in about 30 s.
Overexposure effects	No data available

Safety

Protective gloves	Wear chemical protective gloves.
Eye protection	Wear chemical goggles; wear a mask/respirator in open areas.
Other	Wear additional protective clothing, such as gloves and lab coat with an M9, M17, or M40 mask readily available in closed or confined spaces.
Emergency procedures	**Inhalation:** Remove victim to fresh air; wear a mask/respirator in spite of coughing, sneezing, salivation, and nausea; lift the mask from the face briefly, if necessary, to permit vomiting or to drain saliva from the facepiece; seek medical attention immediately.
	Eye Contact: Don a respiratory protective mask; seek medical attention immediately.
	Skin Contact: Rinse the nose and throat with saline water or bicarbonate of soda solution; wash exposed skin and scalp with soap and water and allow to dry on the skin; dust the skin with borated talcum.

Ingestion: Seek medical attention immediately; carry on duties as vigorously as possible; this will help to lessen and shorten the symptoms; combat duties usually can be performed in spite of the effects of sternutators.

Military Significant Information

Field protection	Protective mask.
Decontamination	None required in the filed. Use alkali solution or DS2 for decontamination in enclosed places.
Persistency	Short, because the compound is disseminated as an aerosol.
Use	Not authorized for US military use.

Chapter 8

Binary Components

GB2 and VX2 are the designations for Sarin (GB) and agent VX that are formed in binary reactions. GB2 and VX2 have been developed to decrease hazards of manufacturing, storing, and handling unitary nerve agents. In binary weapons, two relatively nontoxic chemicals are mixed in flight to form the agent. Compounds used to produce the binary nerve agents are not chemical agents themselves. Chapter 3 discusses these compounds.

The components of GB2 are methylphosphonic difluoride (DF) and a mixture of isopropyl alcohol and isopropyl amine (OPA).

The components of VX2 are QL, which is designated chemically as O-(2-diisopropylaminoethyl)-O'-ethyl methylphosphonite, and NE, which is powdered sulfur with a small amount of added silica aerogel prevent caking.

GB2 is formed by the reaction of DF with OPA.

VX2, binary VX, is formed by the action of O,O'-ethyl (2-diisopropylaminoethyl) methylphosphonite (QL) with sulfur (NE and NM).

DF and its precursor, methylphosphonic dichloride (DC), are organophosphonic acids. They will react with alcohols to form crude lethal nerve agents, such as crude GB. High overexposure may cause inhibition of cholinesterase activity. Although much less toxic than GB, DF and DC are toxic and corrosive materials.

Troop exposure to these materials could result from leaking DF containers, accidents that disrupt packaging, spills at production or storage facilities, or accidents during transport. Because DF and DC are relatively volatile compounds, the primary route of exposure is expected to be the respiratory system. However, ingestion also results from inhalation exposures in animals and could occur in humans. DF and DC vapors have a pungent odor and may cause severe and painful irritation of the eyes, nose, throat, and lungs. Data provided are for DF only, DC has similar properties.

8.1. Alcohol – Amine Mixture – OPA

The mixture known as OPA is relatively nontoxic compared with nerve agents. However, it is not without hazard. It is a highly volatile and flammable liquid composed of 72% isopropyl alcohol and 28% isopropylamine. It forms toxic oxides of nitrogen as well as explosive mixtures in air. In contact with skin and eyes OPA may cause severe irritation. Ingestion causes nausea, salivation, and severe irritation of the mouth and stomach. Inhalation may cause irritation of the lower respiratory tract, coughing, difficult breathing, or loss of consciousness.

Informational

Designation	OPA
Class	Binary component
Type	Does not apply
Chemical name	2-Propanol; isopropylamine
CAS number	Mixture: [not available]; 2-propanol: [67-63-0]; isopropylamine: [75-31-0]

Chemical and Physical Properties

Appearance	Clear liquid
Odor	Alcohol and ammonia
Chemical formula	C_3H_8O; C_3H_9N
Molecular weight	60.1; 59.11
Chemical structure	

$$CH_3-\underset{\underset{H}{|}}{\overset{\overset{OH}{|}}{C}}-CH_3 \ \& \ CH_3-\underset{\underset{H}{|}}{\overset{\overset{NH_2}{|}}{C}}-CH_3$$

Melting point	$< -88°C$
Boiling point	$60°C$
Flash point	$-9.4°C$ (based on amine)
Decomposition temperature	No data available
Vapor density	2.1 (air = 1)
Liquid density	0.7443 g/cm^3 at 25°C
Solid density	No data available
Vapor pressure	197 mmHg at 20°C
Volatility	No data available
Solubility	No data available

Reactivity

Hydrolysis products	No data available
Rate of hydrolysis	No data available
Stability	Stable but reactive, volatile, and flammable
Storage stability	Store OPA in a cool, well-ventilated area away from heat, open flame. Avoid contact with water mists or sprays, metals, alkaline materials, and some organics. Reacts readily with DF or DC, producing extremely toxic compounds (GB or a chlorine compound similar to GB).
Decomposition	No data available
Polymerization	No data available

Toxicity

LD_{50} *(skin)*	No data available
LCt_{50} *(respiratory)*	No data available
LCt_{50} *(percutaneous)*	No data available
ICt_{50} *(respiratory)*	No data available
ICt_{50} *(percutaneous)*	No data available
Rate of detoxification	No data available
Skin and eye toxicity	Irritating and toxic through the skin and eyes, mainly because of the isopropylamine in the mixture.
Rate of action	No data available
Overexposure effects	No data available

Safety

Protective gloves	Wear appropriate protective gloves to prevent any possibility of contact with skin; butyl and neoprene rubber gloves are preferred.
Eye protection	Wear splash-proof or dust-resistant safety goggles and a faceshield to prevent contact with substance.
Other	Wear respirators based on contamination levels found, must not exceed the working limits of the respirator and must be jointly approved by NIOSH; employer should provide an eye wash fountain and quick drench shower for emergency use.
Emergency procedures	**Inhalation:** If adverse effects occur, remove to uncontaminated area. Give artificial respiration if not breathing. If breathing is difficult, oxygen should be administered. Get immediate medical attention.

Eye Contact: Immediately flush eyes with plenty of water for at least 15 min. Then get immediate medical attention.

Skin Contact: Wash skin with soap and water for at least 15 min while removing contaminated clothing and shoes. Get immediate medical attention. Thoroughly clean and dry contaminated clothing and shoes before reuse. Destroy contaminated shoes.

Ingestion: Contact local poison control center or physician immediately. Never make an unconscious person vomit or drink fluids. Give large amounts of water or milk. Allow vomiting to occur. When vomiting occurs, keep head lower than hips to help prevent aspiration. If person is unconscious, turn head to side. Get medical attention immediately.

Military Significant Information

Field protection	Protective mask, and clothing
Decontamination	Large volumes of water to flush OPA from the skin. Wash clothing with water. Absorb spilled OPA with vermiculite, earth, or sand.
Persistency	No data available
Use	Binary GB component

8.2. Binary Component – QL

One of the binary components for VX is QL, an organophosphorous ester. An additional designation for QL is EDMP, an abbreviation for *O,O'*-ethyl (2-diisopropylaminoethyl) methylphosphonite. The pure material is many times less toxic than VX, but is by no means harmless. It reacts with moisture and other substances to produce highly toxic materials as well as flammable materials. It will ignite without application of spark or flame at 129°C (265°F). A hydrolysis product of QL ignites at a much lower temperature. QL is a slight cholinesterase inhibitor, but the body tissue and fluids do not store it for extended periods. Prolonged breathing of QL vapors may produce headaches and nausea.

Informational

Designation	QL
Class	Binary component
Type	Does not apply
Chemical name	*O'*-ethyl-*O*-(2-diisopropylaminoethyl) methylphosphonite
CAS number	[57856-11-80]

Chemical and Physical Properties

Appearance	Viscous liquid
Odor	Strong fishy
Chemical formula	$C_{11}H_{26}NOP$
Molecular weight	219.31

Chemical structure

$$CH_3CH_2 - \overset{\overset{\displaystyle CH_3}{\displaystyle |}}{P} - OCH_2CH_2N[CH(CH_3)_2]_2$$

Melting point	No data available
Boiling point	232°C
Flash point	89°C, in addition QL has an autoignition temperature of 129°C
Decomposition temperature	No data available
Vapor density	8.1 (air = 1)
Liquid density	0.908 g/cm³ at 25°C
Solid density	No data available
Vapor pressure	0.01 mmHg at 20°C
Volatility	No data available
Solubility	No data available

Reactivity

Hydrolysis products	With an excess of water by weight, QL form 2-diisopropylaminoethanol, ethanol, and methyl phosphorous acid. With traces of water or other proton donors QL will produce diethylmethyl phosphonite and *O,O'*-bis-(2-diisopropylaminoethyl) methylphosphonite. Diethylmethyl phosphonite has a boiling point of 120°C and a vapor pressure of 11 mmHg at 20°C and is highly flammable.
Rate of hydrolysis	Rapid. QL can be hydrolyzed and oxidized.
Stability	Unstable in air. Protect from water or moisture. Store away from heat or ignition sources and sulfur compounds. Reacts with sulfur and sulfur compounds, producing highly toxic VX or VX-like compounds. It completely dissolves poly-methylmethacrylate. It is incompatible with calcium hypochlorite (HTH), many chlorinated hydrocarbons, selenium, selenium compounds, moisture, oxidants, and carbon tetrachloride.
Storage stability	No data available

Decomposition	No data available
Polymerization	No data available

Toxicity

LD$_{50}$ (skin)	No data available
LCt$_{50}$ (respiratory)	No data available
LCt$_{50}$ (percutaneous)	No data available
ICt$_{50}$ (respiratory)	No data available
ICt$_{50}$ (percutaneous)	No data available
Rate of detoxification	No data available
Skin and eye toxicity	Skin irritant but not a skin sensitizer or eye irritant. Animal studies have shown QL to be relatively low toxic, although the hydrolysis products are highly toxic.
Rate of action	No data available
Overexposure effects	No data available

Safety

Protective gloves	Wear appropriate protective gloves to prevent any possibility of contact with skin; butyl and neoprene rubber gloves are preferred.
Eye protection	Wear splash-proof or dust-resistant safety goggles and a faceshield to prevent contact with substance.
Other	Wear respirators based on contamination levels found, must not exceed the working limits of the respirator and must be jointly approved by NIOSH; employer should provide an eye wash fountain and quick drench shower for emergency use.
Emergency procedures	***Inhalation:*** If adverse effects occur, remove to uncontaminated area. Give artificial respiration if not breathing. If breathing is difficult, oxygen should be administered. Get immediate medical attention.
	Eye Contact: Immediately flush eyes with plenty of water for at least 15 min. Then get immediate medical attention.
	Skin Contact: Wash skin with soap and water for at least 15 min while removing contaminated clothing and shoes. Get immediate medical attention. Thoroughly clean and dry contaminated clothing and shoes before reuse. Destroy contaminated shoes.
	Ingestion: Contact local poison control center or physician immediately. Never make an unconscious person vomit or drink fluids. Give large amounts of water or milk. Allow

vomiting to occur. When vomiting occurs, keep head lower than hips to help prevent aspiration. If person is unconscious, turn head to side. Get medical attention immediately.

Military Significant Information

Field protection	Protective mask, gloves, and clothing if severe exposure exists.
Decontamination	Soda ash, slaked lime, limestone, or sodium bicarbonate. Cover spills with vermiculite, diatomaceous earth, clay, or fine sand followed by one of the decontaminants.
Persistency	No data available
Use	Binary VX component

8.3. Difluoro – DF

DF and its precursor, DC are organophosphonic acids. They will react with alcohols to form crude lethal nerve agents, such as crude GB. High overexposure may cause inhibition of cholinesterase activity. Although much less toxic than GB, DF and DC are toxic and corrosive materials. Because DF and DC are relatively volatile compounds, the primary route of exposure is expected to be the respiratory system. However, ingestion also results from inhalation exposures in animals and could occur in humans. DF and DC vapors have a pungent odor and may cause severe and painful irritation of the eyes, nose, throat, and lungs. Data provided is for DF only, DC has similar properties.

Informational

Designation	DF
Class	Binary component
Type	Does not apply
Chemical name	Methylphosphonic difluoride
CAS number	[676-99-3]

Chemical and Physical Properties

Appearance	Liquid
Odor	Pungent, acid-like
Chemical formula	CH_3F_2PO
Molecular weight	100.1
Chemical structure	$$CH_3 - \overset{\displaystyle O}{\overset{\displaystyle \|}{\underset{\displaystyle \underset{\displaystyle F}{\|}}{P}}} - F$$

Melting point	−37.1°C
Boiling point	100°C
Flash point	Does not flash
Decomposition temperature	No data available
Vapor density	3.45 (air = 1)
Liquid density	1.359 g/cm^3 at 25°C
Solid density	No data available
Vapor pressure	36 mmHg at 20°C
Volatility	147,926 mg/m^3 at 19.5°C
Solubility	No data available

Reactivity

Hydrolysis products	Hydrolyzes to give toxic products, MF and HF. Further hydrolysis of MF results in methylphosphonic acid.
Rate of hydrolysis	Virtually instantaneous to products (MF and HF) that are toxic. Further hydrolysis is a slow reaction. Half-life for hydrolysis of MF is 162 days at pH 7; 90 days at pH 4; and 47 days at pH 3.
Stability	Does not spontaneously decomposes but is reactive. Avoid contact with water mists or sprays, metals, alkaline materials, and some organics.
Storage stability	Store DF in lead and wax-lined carboys, high-density polyethylene bottles, or nickel-lined containers in well-ventilated areas. Never store DF with alcohols; DF will react with alcohols to form lethal chemicals, such as crude GB. Incompatible with water, glass, concrete, most metals, natural rubber, leather, and organic materials like glycols. The acidic corrosive hydrolysis products may react with metals, such as Al, Pb, and Fe, to give off hydrogen gas, a potential fire and explosive hazard.
Decomposition	No data available
Polymerization	No data available

Toxicity

LD$_{50}$ (skin)	No data available
LCt$_{50}$ (respiratory)	No data available
LCt$_{50}$ (percutaneous)	No data available
ICt$_{50}$ (respiratory)	No data available
ICt$_{50}$ (percutaneous)	No data available
Rate of detoxification	No data available

Skin and eye toxicity	May cause severe and painful irritation of the eyes, nose, throat, and lungs. DF hydrolyzes to HF, which may cause second- or third-degree burns upon contact.
Rate of action	No data available
Overexposure effects	No data available

Safety

Protective gloves	Wear appropriate protective gloves to prevent any possibility of contact with skin; butyl and neoprene rubber gloves are preferred.
Eye protection	Wear splash-proof or dust-resistant safety goggles and a faceshield to prevent contact with substance.
Other	Wear respirators based on contamination levels found, must not exceed the working limits of the respirator and must be jointly approved by NIOSH; employer should provide an eye wash fountain and quick drench shower for emergency use.
Emergency procedures	***Inhalation:*** If adverse effects occur, remove to uncontaminated area. Give artificial respiration if not breathing. If breathing is difficult, oxygen should be administered. Get immediate medical attention. ***Eye Contact:*** Immediately flush eyes with plenty of water for at least 15 min. Then get immediate medical attention. ***Skin Contact:*** Wash skin with soap and water for at least 15 min while removing contaminated clothing and shoes. Get immediate medical attention. Thoroughly clean and dry contaminated clothing and shoes before reuse. Destroy contaminated shoes. ***Ingestion:*** Contact local poison control center or physician immediately. Never make an unconscious person vomit or drink fluids. Give large amounts of water or milk. Allow vomiting to occur. When vomiting occurs, keep head lower than hips to help prevent aspiration. If person is unconscious, turn head to side. Get medical attention immediately.

Military Significant Information

Field protection	Protective mask, gloves, and clothing
Decontamination	Water. Flush eyes for 15 min, seek medical attention.
Persistency	No data available
Use	Binary GB component

Appendix A

Glossary

A

AC – hydrogen cyanide.

AChE – acetylcholinesterase.

acetylcholine – a chemical neuro-transmitter produced by nerve cells predominantly outside the CNS. It is a chemical "messenger," stimulating the heart, skeletal muscles, and numerous secretory glands.

acetylcholinesterase – an enzyme that normally hydrolyzes acetyl-choline, thereby stopping its activity.
  Acetylcholinesterase is inhibited by organophosphates, carbamates, and glycolates.

acid – a chemical compound having a pH less than 7. Acids usually have a sour taste and a propensity to react with bases to form salts.

acute – having a short and relatively severe course; arising quickly, as acute symptoms.

aerosol – a liquid or solid composed of finely divided particles suspended in a gaseous medium. Examples of common aerosols are mist, fog, and smoke.

algogen – a substance that produces pain.

alkali – a class of bases that neutralize acids and form salts. Sodium hydroxide (lye) and ammonium hydroxide are alkalies.

alkaline – having the properties of an alkali, for example, sodium hydroxide; opposed to acid. Having hydroxyl ions (OH); basic.

alkaloids – a group of basic organic substances of plant origin. Many have important physiological actions and are used in medicine, for example, cocaine.

amino acids – basic building block units that can be chemically linked together to form larger molecules, such as peptides and proteins.

analgesic – substance used in medicine to relieve pain.

analogue – a chemical compound similar in structure to another chemical compound and having the same effect on body functions.

antiplant – herbicide.

aqueous – watery; prepared with water.

arrhythmia – any variation from the normal rhythm of the heartbeat.

arsenical – a chemical compound containing arsenic.

atropine – an alkaloid obtained from *Atropa belladonna*. It is used as an antidote for nerve agent poisoning. It inhibits the action of acetylcholine at the muscle junction by binding to acetylcholine receptors.

autonomic nervous system – that part of the nervous system that governs involuntary functions, such as heart rate, reflexes, and breathing. It consists of the sympathetic and parasympathetic nervous systems.

B

basic – relating to a base; having a pH greater than 7.

binary chemical munition – a munition in which chemical substances, held in separate containers, react when mixed or combined as a result of being fired, launched, or otherwise initiated to produce a chemical or antimateriel agent.

binary components – the component chemicals that combine to produce binary chemical agents. Examples of two common binary chemical agent components are as follows:

a. The components for binary GB (GB2) are methylphosphonic difluoride (DF) and isopropyl alcohol with an amine added (OPA).

b. The components for binary VX (VX2) are ethyl 2-di-isopropylaminoethyl methyl phosphonite (QL) and dimethylpolysulfide (NM).

biological agent – a microorganism that causes disease in people, plants, or animals or causes the deterioration of materiel.

biological operation – employment of biological agents to produce casualties in man or animal and damage to plants or materiel; or defense against such employment.

biological warfare – see biological operation.

botulism – poisoning by toxin derived from the microorganism *Clostridium botulinum*.

BW – biological warfare.

BZ – a CNS depressant.

C

C – average concentration of an agent in the atmosphere; Celsius.

CA – bromobenzylcyanide.

carbamates – organic chemical compounds that can be neurotoxic by competitively inhibiting acetylcholinesterase binding to acetylcholine.

CAS – Chemical Abstracts Service.

casualty – any person who is lost to the organization by reason of having been declared dead, wounded, injured, diseased, interned, captured, retained, missing, missing in action, beleaguered, besieged, or detained.

catalyst – a material that increases or decreases the rate of a chemical reaction without being changed by the reaction.

central nervous system (CNS) – consists of the brain and spinal cord. The CNS controls mental activity and voluntary muscular activity. It also coordinates the body's involuntary functions indirectly.

CG – phosgene.

chemical agent – a chemical substance that is intended for use in military operations to kill, seriously injure, or incapacitate people through its physiological effects. Excluded from consideration are riot control agents, chemical herbicides, and smoke and flame materials. Included are blood, nerve, choking, blister, and incapacitating agents.

chemical agent casualty – a person who has been affected sufficiently by a chemical agent to prevent or seriously degrade his or her ability to carry out the mission.

chemical agent symbol – the military Army code designation of any chemical agent. This is a combination of one to three letters or letter and number combinations. Do not confuse the symbol with the chemical formula.

chemical contamination – the presence of an agent on a person, object, or area. Contamination density of an agent is usually expressed either in milligrams or grams per square meter (mg/m}, g/m}) or in pounds per hectare (lb/ha). A hectare is 10,000 m².

CK – cyanogen chloride.

CL – chlorine

CMPF – cyclohexyl methyl phosphonofluoridate

CN – chloroacetophenone.

CNOH – cyanic acid.

CNS – central nervous system.

CO – carbon monoxide.

compound – in chemical terms, a uniform substance formed by the stable combination of two or more chemical elements, as distinct from a mixture.

concentration – the amount of an agent present in a unit volume. Usually expressed in milligrams per cubic meter (mg/m³) of air.

contaminate – to introduce an impurity; for instance, foreign microorganisms developing accidentally in a pure culture. Clothing containing microorganisms is said to be contaminated.

coronary – pertaining to the heart.

covert – hidden, concealed, insidious.

CR – dibenz-(b,f)-l,4-oxazepine.

CS – O-chlorobenzylidene malononitrile, a tear agent.

Ct – vapor dosage.

cutaneous – pertaining to the skin.

CX – phosgene oxime.

cyanosis – blueness of the skin owing to insufficient oxygen in the blood.

cytotoxin – toxin that directly damages and kills the cell with which it makes contact.

D

DA – diphenylchloroarsine, a vomiting agent.

DC – diphenylcyanoarsine, a vomiting agent.

decay rate – the predictable rate at which microorganisms die.

decontaminating material – any substance used to destroy chemically or by other means, to physically remove, seal, or otherwise make the agents harmless.

decontamination – the process of making any person, object, or area safe by absorbing, destroying, neutralizing, making harmless, or removing chemical or biological agents, or by removing radioactive material clinging to or around it.

defoliant – an agent that, when applied to plants, kills or damages them or causes them to shed their leaves.

dehydrate – to remove water from.

depolarize – to remove the polarity, or difference in electrical charge, as on opposite sides of a cell membrane. When a nerve or muscle cell is stimulated it becomes depolarized.

desiccant – a substance that has an affinity (attraction) for water. When used as defoliants, desiccants remove water from plant tissue causing it to dry and shrivel.

desiccate – to dry completely.

detection – the determination of the presence of an agent.

detoxification rate – rate at which the body's own actions overcome or neutralize (detoxify) chemicals or toxins. Agents that the body cannot easily break down and neutralize and that accumulate in the body are called "cumulative."

DF – methylphosphonic difluoride.

dilate – to make wider or larger.

dilute solution – chemical agents that have been reduced in strength by dilution.

disease – a deviation from the normal state of function of a cell, an organ, or an individual.

disinfect – to free from pathogenic organisms or to destroy them.

disinfectant – an agent, usually chemical, that destroys infective agents.

dissemination – distribution or spreading.

DM – diphenylaminochloroarsine (Adamsite), a vomiting agent.

DMSO – dimethylsulfoxide.

DNA – deoxyribonucleic acid.

dosage – cumulative exposure equivalent to the concentration of chemical agent to which an individual is exposed integrated over the time of exposure.

dose – quantity of agent having entered the body.

DP – diphosgene.

DS2 – decontaminating solution number 2.

dyspnea – difficult or labored breathing.

E

ECt – effective dosage of an aerosol.

ED – ethyldichloroarsine.

edema – excessive accumulation of fluid in body tissues or body cavities.

EDMP – O, O'-ethyl (2-diisopropylaminoethyl) methylphosphonite, one of the binary components of VX.

endemic – native to, or prevalent in, a particular district or region. An endemic disease has a low incidence, but is constantly present in a given community.

endogenous – produced or originating from within.

endogenous biological regulators – naturally occurring, biological regulators with potential for chemical and biological warfare applications.

environment – the external surroundings and influences.

enzyme – organic substance capable of causing chemical changes to take place quickly at body temperature by catabolic action as in digestion.

eruption – a rash, visible lesion, or injury of the skin characterized by redness, prominence, or both.

eutectic mixture – a mixture of two or more substances in proportions that give the lowest freezing or melting point. The minimum freezing point attainable is termed the eutectic point.

exotoxin – a toxin excreted by a microorganism into the surrounding medium.

F

fatigue – weariness from labor or exhausting conditions where cells or organs have undergone excess activity so that they respond to stimulation with less than normal activity.

FDF – fast death factor.

fever – abnormally high body temperature; characterized by marked increase of temperature, acceleration of the pulse, increased tissue destruction, restlessness, and sometimes delirium.

flame – burning gas or vapor that causes lethal or incapacitating effects by means of direct burn wounds, depletion of oxygen, carbon monoxide poisoning, heat, or a combination of these factors. Flame can function secondarily as an incendiary.

flash point – the lowest temperature at which a substance gives off enough combustible vapors to produce momentary ignition when a flame is applied under controlled conditions.

G

GA – tabun.

ganglia – a knot-like mass of neurons located outside the CNS.

GB – sarin.

GB2 – a binary nerve agent.

GD – soman.

GE – ethyl sarin.

GF – cyclo sarin.

gene – a segment of a chromosome definable in operational terms as a unit of genetic (inheritable) information.

G-series nerve agents – a series of nerve agents developed by the Germans, that includes Tabun (GA), Sarin (GB), Soman (GD), Ethyl Sarin (GE).

GV – a persistent nerve agent.

H

H – Levinstein mustard, a blister agent.

H-series agents – a series of persistent blister agents, that include distilled mustard (HD) and the nitrogen mustards (HN-1, HN-2, and HN-3).

half-life – time required for half a material to decompose.

harassing concentration – a concentration of an agent that requires masking or other protective measures. Such concentration may be insufficient to kill, but sufficient to interfere with normal operations.

HCl – hydrogen chloride.

HCN – hydrogen cyanide.

HD – distilled mustard, a blister agent.

hemolysis – the destruction of red blood cells followed by release of the hemoglobin they contain.

hemorrhage – bleeding.

hepatitis – inflammation of the liver.

herbicide – a chemical compound that will kill or damage plants.

HF – hydrogen fluoride.

Hg – mercury.

HL – mustard–lewisite mixture.

HN-1 – nitrogen mustard 1.

HN-2 – nitrogen mustard 2.

HN-3 – nitrogen mustard 3.

HT – mustard–T mixture.

HTH – calcium hypochlorite.

hydrolysis – interaction of a chemical agent with water to yield a less toxic product or products.

hydrolyze – to subject to hydrolysis; to split a chemical bond with water.

hygiene – the science of health and the preservation of good health.

I

ICt_{50} – median incapacitating dosage of a chemical agent vapor or aerosol.

ID_{50} – median incapacitating dosage of a liquid chemical agent.

identification – can be subdivided into the following two levels:

Definitive identification – the determination of the exact identity of a compound or organism through the establishment of a group of unique characteristics.

Classification – the determination that a compound or organism is a member of a chemical or biological class without knowledge of the exact identity of the compound or organism.

incapacitation – disablement.

incendiary – primarily an antimateriel compound that generates sufficient heat to cause destructive thermal degradation or destructive combustion of materiel.

industrial chemicals – chemicals developed or manufactured for use in industrial operations or research, by industry, government, or academia. These chemicals are not primarily manufactured for the specific purpose of producing human casualties or rendering equipment, facilities, or areas dangerous for use by man. Hydrogen cyanide, cyanogen chloride, phosgene, and chloropicrin are industrial chemicals that also can be military chemical agents (AC, CK, CG, and PS).

inflammation – reaction of tissues to injury; characterized by pain, heat, redness, or swelling of the affected parts.

intoxication – poisoning.

intracellular – inside or within the cell.

intraperitoneal – within the abdominal cavity.

intravenous – within the vein.

ion – an atom that has acquired an electrical charge because of gain or loss of electrons.

ion-channel – a passage that allows particular charged particles, such as sodium ions, potassium ions, or calcium ions, to pass through a membrane. Ions do not cross cell membranes through simple pores.

ionophore – a substance which creates a passage through membranes for ions.

ip – intraperitoneal.

J

jaundice – a disease symptom characterized by yellowing of the skin and eyes and by a deep yellow color of the urine.

K

K agents – incapacitating agents.

kg – kilogram(s).

L

L-1 – lewisite 1.

L-2 – lewisite 2.

L-3 – lewisite 3.

lacrimator – a compound that causes a large flow of tears and irritates the skin.

latent period – a period of seeming inactivity.

LCt_{50} – median lethal dosage of a chemical agent vapor or aerosol.

LD_{50} – median lethal dosage of a liquid chemical agent.

lesion – injury, diseased area or pathological change in an organ or tissue.

lethal chemical agent – an agent that maybe used effectively in field concentrations to kill.

M

malaise – a feeling of bodily discomfort.

malignant – tending to go from bad to worse; capable of spreading from one site within the tissues to another.

MD – methyldichloroarsine.

membrane – a thin layer of tissue that covers a surface or divides a space or organ.

MF – methylphosphonofluoridic acid.

mg – milligram(s).

mg/kg – milligram(s) per kilogram (of body weight).

mg-min/m^3 – milligram-minute(s) per cubic meter.

microencapsulate – to make, form, or place in an extremely small or microscopic capsule.

micron – a unit of measurement: 1/1000 mm. Usually designated by the Greek letter μ.

military chemical compound – chemical substance that has become accepted generally by the public for use in conventional war. Included are riot control agents, smoke and flame materials, and military herbicides. Excluded are chemical agents.

miosis – excessive contraction of the pupil.

miscible – capable of being mixed.

mm – millimeter(s).

mm Hg – millimeters of mercury; a unit used to describe atmospheric pressure.

molecular weight – sum of the atomic weights of each atom in a molecule.

molecule – a chemical combination of two or more atoms that form a specific chemical substance.

monitoring – the act of detecting the presence of radiation and the measurement thereof with radiation measuring instruments.

MOPP – mission-oriented protective posture, level 4.

mortality rate – the ratio of the number of deaths from a given disease to the total number of cases of that disease.

MPOD – another designation for DF and DC.

mw – molecular weight.

N

N – normality.

NaOH – sodium hydroxide.

nausea – tendency to vomit; sickness of the stomach.

NE – powdered elemental sulfur mixture.

neat chemical agent – a nondiluted, full-strength (as manufactured) chemical agent. A chemical agent manufactured by the binary synthesis route will also be considered a neat agent regardless of purity.

necrosis – death of a cell or group of cells.

necrotic – capable of destroying living tissue.

neural – relating or pertaining to nerves.

neuron – a nerve cell. Neurons are characterized by their ability to become excited and to transmit their excitation onto other cells.

neurotoxic – poisonous to nerve tissue.

neuropeptide – a peptide produced by certain nerve cells, particularly in the CNS. Some may act as neurotransmitters or neurohormones. Acetylcholine, norepinephrine, serotonin, and histamine are neuropeptides.

neurotransmitters – chemical substances released by neurons into the synapse and causing an effect on the postsynaptic cell. More than 50 compounds have been identified as neurotransmitters, including acetylcholine.

neutralize – to render neutral.

ng – nanogram(s).

NIOSH – National Institute of Occupational Safety and Health.

NM – dimethyl–polysulfide mixture.

NO – nitric oxide.

normality – in chemistry, a solution concentration designated by the number of gram equivalents of solute per liter of solution.

O

OPA – isopropylamine and isopropyl alcohol.

organic solvent – an organic chemical compound that dissolves another to form a solution. Examples of organic solvents are alcohols, turpentine, kerosene, benzene, chloroform, acetone, carbon tetrachloride, and toluene. Degreasers, paint thinners, antifreeze, and dry-cleaning compounds contain organic solvents.

organophosphate – a phosphate-containing organic compound. Organophosphates inhibit cholinesterase enzymes. G-series and V-series nerve agents are organophosphates, as are certain common insecticides.

oxidative processes – chemical reactions requiring oxygen.

oxime – a chemical compound containing one or more oxime groups (NOH). Although the chloroformoximes are blister agents, some oximes are beneficial. 2-PAM chloride (trade name protopam chloride or pralidoxime chloride) is used in treatment of nerve agent poisoning. This drug increases the effectiveness of drug therapy in poisoning by some, but not all, cholinesterase inhibitors (nerve agents). It reactivates the inhibited enzyme at skeletal muscles as well as at parasympathetic sites (glands and intestinal tract) and therefore relieves the skeletal neuromuscular block that causes the paralysis associated with the nerve agents.

P

2-PAM chloride – 2-pralidoxime chloride. See oxime.

parasympathetic nervous system – the part of the autonomic nervous system that decreases pupil size, heart rate, and blood pressure and increases functions, such as secretion of saliva, tears, and perspiration.

parenteral – in some manner other than by the intestinal tract.

passive immunity – immunity acquired by introduction of antibodies produced in the body of another individual animal.

pathogen – a disease-producing microorganism.

pathogenic – causing disease.

PB – pyridostigmine bromide.

PD – phenyldichloroarsine.

peptide – an organic compound of amino acids linked together by peptide bonds.

percutaneous – effected or performed through the skin.

persistency – in biological or chemical warfare, the characteristics of an agent which pertains to the duration of its effectiveness under determined conditions after its dispersal.

PFIB – perfluoroisobutylene.

pH – the chemist's measure of acidity and alkalinity. It is a scale in which the number 7 indicates neutral; numbers below 7 indicate acidity; and numbers above 7 indicate an alkaline substance.

physostigmine – an alkaloid from the calabar bean *Physostigma*. Physostigmine salicylate is used to relieve symptoms of BZ and other glycolate exposure.

polypeptide – a polymer of numerous amino acid residues (usually more than 20) linked together chemically by peptide bonds.

postmortem – occurring or performed after death.

postsynaptic – after a synapse.

potable – fit or suitable for drinking.

potassium – a chemical element important along with sodium in maintaining cell volume and an electrical gradient across cell membranes. Like sodium, potassium is important in nerve and muscle function.

presynaptic – before a synapse.

prostration – extreme exhaustion.

proteins – a class of organic compounds of very high molecular weights which compose a large part of all living matter.

PS – chloropicrin, a vomiting agent.

pulmonary – pertaining to the lungs.

pyridostigmine bromide – an antidote enhancer that blocks acetylcholinesterase, protecting it from nerve agents. When taken in advance of nerve agent exposure, PB increases survival provided that atropine and oxime (Mark I NAAK) and other measures are taken.

Q

QL (EDMP) – an organophosphorous ester, one of the components of VX.

R

RCA – riot control agent.

reagent – a chemical substance used to produce a chemical reaction.

receptor – a component of cell membranes where specific compounds bind, causing a change in the biological activity of the cell. Cells have receptors that can bind neurotransmitters, toxins, viruses, and other agents.

respiration – the act or function of using oxygen.

respiratory – pertaining to respiration.

S

SA – arsine.

slurry – a thin, watery mixture.

sodium – one of the two chemical elements in table salt (the other is chlorine). In the body, sodium is one of the most important constituents of blood and other body fluids. Its balance inside and outside cells is important in proper cell function including nerve and muscle activity.

specific gravity – the weight of a particular volume of a substance compared with the weight of an equal volume of water.

STB – supertropical bleach.

sternutator – vomiting compound.

stupor – partial or nearly complete unconsciousness.

survey – the directed effort to determine the location and the nature of the agent in an area.

suspension – a mixture of fine particles and a liquid. If the mixture is allowed to stand, the fine particles will settle.

sympathetic nervous system – a network of nerves that trigger certain involuntary and automatic bodily functions, such as constricting blood vessels, widening the pupils, and speeding up the heartbeat.

symptoms – functional evidence of disease; a change in condition indicative of some mental or bodily state.

synapse – site at which neutrons make functional contacts with other neurons or cells.

synergistic – working together; having combined cooperative action that increases the effectiveness of one or more of the components' properties.

synthesize – to build up a chemical compound from its elements or other compounds.

systemic – relating to the entire organism instead of a part.

systemic action – action affecting many systems. It includes the movement of the agent through the organism and its effect on cells and processes remote from the point of application.

T

t – time.

TGD – thickened Soman.

thickened agent – an agent to which a polymer or plastic has been added to retard evaporation and cause it to adhere to surfaces.

TOF – trioctylphosphite.

toxemia – a general poisoning or intoxication owing to absorption of products (toxins) of microorganisms formed at a local source of infection.

toxic – poisonous; effects ranging from harmful to lethal depending on the dose and resistance of the individual.

toxicity – the quality of being poisonous.

training agent – an agent authorized for use in training to enhance proficiency for operating under NBC conditions.

T-2 – trichothecene, a mycotoxin.

TR – O,O'-diethylmethylphosphonite.

U

μg – microgram(s).

ultraviolet – light waves shorter than the visible blue-violet waves but longer than X-rays. Ultraviolet radiation is very effective in killing unprotected microorganisms.

urticant – a substance which produces a stinging sensation, as if with nettles. Phosgene oxime (CX) is an urticant.

UV – ultraviolet.

V

V-agents – persistent, highly toxic nerve agents developed in the mid-1950s and absorbed primarily through the skin.

vapor pressure – the pressure exerted at any temperature by a vapor when a state of equilibrium has been reached between it and its liquid or solid state.

vesicant – agent that acts on the eyes and lungs, capable of producing blisters, and blisters the skin.

VE – a persistent nerve agent.

VFDF – very fast death factor.

VG – amiton.

viable – capable of living.

viscosity – the resistance of a liquid to flow, resulting from the combined effects of internal friction and friction between the liquid and its surroundings.

viscous – resisting flow.

VM – edemo.

volatile – passing readily into a vapor; having a high vapor pressure.

volatility – the tendency of a chemical to vaporize or give off fumes.

It is directly related to vapor pressure.

VR – Russian VX.

VS – a persistent nerve agent.

VX – a persistent nerve agent, the US standard V-agent.

Vx – a persistent nerve agent.

W

weapon system – an integrated relationship of agents, munitions, or spraying devices and their mode of delivery to the target.

WP – white phosphorous.

Z

Zone I – an area of major operational concern in the predicted biological downwind hazard area. Casualties may exceed 30% in unprotected personnel.

Appendix B

Chemical and Physical Concepts

Boiling Point

Boiling point is the temperature at which the vapor pressure of a liquid equals the atmospheric pressure. The boiling point represents the highest usable temperature of a liquid agent. You can use it to estimate the persistency of a chemical (under a given set of conditions). The reason is that the vapor pressure and the evaporating tendency of a chemical agent vary inversely with its boiling point.

The higher the boiling point, the more slowly a liquid evaporates at ordinary temperatures. For example, HD boils at 217°C and evaporates relatively slowly at ordinary temperatures. CG boils at 7.5°C and evaporates rapidly at moderate temperatures. Thus, agents with low boiling points are normally nonpersistent. Those with high boiling points are persistent. The boiling point also gives an indication of the practicality for decontamination with hot air.

Decomposition Temperature

The decomposition temperature is that at which a chemical breaks down into two or more substances. This temperature can be used to evaluate candidate chemical agents. A low decomposition temperature (one that is markedly lower than the boiling point) will usually mean that dissemination of the chemical agent will cause excessive decomposition.

Flash Point

The flash point is the temperature at which a chemical agent gives off enough vapors to be combustible upon application of a flame under controlled conditions.

The flash point is of interest with chemical agents that have a low enough flash points to cause them to burn when the containing munition bursts.

Freezing Point

Freezing point is the temperature at which a liquid changes to a solid. It is generally equivalent to the melting point. It is important to know the freezing point of a chemical agent, because dissemination characteristics vary markedly with physical state. For example, HD can freeze in a spray tank at low temperatures and cannot be dispensed.

Below their freezing points, most agents become unreliable in creating casualties by direct effect. The effects result from secondary transfer. Warming the frozen agent upon entering an enclosure or area where the temperature is high enough will melt or vaporize the agent.

Hydrolysis

Hydrolysis is the reaction of a compound with the elements of water whereby decomposition of the substance occurs. The reaction produces one or more new substances.

Rapid hydrolysis aids in lowering the duration of effectiveness of toxic chemical agents. For example, in the presence of water or water vapor, lewisite (L) rapidly hydrolyzes. Therefore, it has a shorter duration of effectiveness than distilled mustard (HD).

New substances (hydrolysis products) form when an agent or compound reacts with water. In certain cases hydrolysis does not completely destroy the toxicity of an agent or compound. The resulting hydrolysis products may also be toxic. Examples include lewisite and other agents containing arsenic.

Liquid and Solid Densities

The density of a liquid chemical agent is the weight in grams of one cubic centimeter of the liquid at a specified temperature. The density of a solid chemical agent is the weight of one cubic centimeter of the solid at a specified temperature. Liquid density in this manual is a measure of an agent's weight in comparison with water (density 1.0). This comparison gives the specific gravity. Liquids form layers, and the one with the greatest density sinks to the bottom. The layer with the lowest density rises to the top.

If an agent is much denser than water, it will tend to sink to the bottom and separate out. An example is mustard lewisite (liquid density 1.66). If an agent has about the same density as water, such as the nerve agents, it will tend to mix throughout the depth of the water. (However, GB is the only nerve agent that will actually dissolve in water.) The nerve agents are slightly denser than water. Therefore, the concentration of nerve agent will tend to increase with depth. Agents that are less dense than water, such as AC (liquid density 0.69), tend to float on water.

Note: Sometimes density is not the only factor that determines how an agent will distribute in water. An example is HD. If agent HD falls on water, little globs of HD will scatter throughout the depth of the water. The larger ones tend to coagulate and sink, forming a layer at the bottom. The finer droplets or mist created by the explosion form a layer on top, one molecule thick, like an oil slick. Surface tension holds the layer on the surface. If the slick is very thin, it will be iridescent, reflecting rainbow-like colors. If it is a little thicker, it may not be noticeable.

Liquid density is of interest in computing the chemical efficiency of a munition, because we always express toxicities in units of weight. For example, a munition filled with CG (liquid density about 1.4 at 20˚C) will contain twice as much chemical agent by weight as a munition of the same volume filled with AC (liquid density about 0.7 at 10˚C). It will also have a much higher chemical efficiency. To find the chemical efficiency of a munition, divide the weight of the filling by the total weight of the filled munition.

Median Lethal Dosage (LD$_{50}$) of Liquid Agent

The LD$_{50}$ is the amount of liquid agent expected to kill 50% of a group of exposed, unprotected personnels.

The ID$_{50}$ is the amount of liquid agent expected to incapacitate 50% of a group of exposed, unprotected individuals.

For airborne chemical agents, the concentration of agent in the air and the time of exposure are the important factors that govern the dose received. The dosage may be inhaled (respiratory) or absorbed through the eyes (ocular) or through the skin (percutaneous). Dosages are based on short exposures – 10 min or less. Toxicity is generally identified by reference to the lethal dosage.

Median Lethal Dosage (LCt$_{50}$) of a Vapor or Aerosol

The median lethal dosage of a chemical agent employed for inhalation as a vapor or aerosol is generally the LCt$_{50}$. The LCt$_{50}$ of a chemical agent is the dosage (vapor concentration of the agent multiplied by the time of exposure) that is lethal to 50% of exposed, unprotected personnel at some given breathing rate. It varies with the degree of protection provided by masks and clothing worn by personnel and by the breathing rate. If individuals are breathing faster, they will inhale more agents in the same time, increasing the dose received.

Median Incapacitating Dosage (ICt$_{50}$) of a Vapor or Aerosol

For inhalation effect, the median incapacitating dosage is the ICt$_{50}$. The ICt$_{50}$ is the amount of inhaled vapor that is sufficient to disable 50% of exposed, unprotected personnel. The unit used to express ICt$_{50}$ is mg-min/m^3.

Note: You may also express dosages in amounts other than the median dosage. For example, the LCt$_{25}$ is the dosage of vapor that would kill 25% of a group of exposed, unprotected personnel; ICt$_{90}$ is the vapor dosage that would incapacitate 90% of a group of exposed, unprotected personnel.

Melting Point

Melting point is the temperature at which a solid changes to a liquid. White phosphorous (WP) presents an example of the importance of melting point. It has a low melting point. In temperatures above that melting point, you must store any WP-filled munition on its end. When the WP melts, the center of gravity will remain unchanged and thus prevent instability of the munition in flight.

Molecular Weight

Molecular weight (MW) is the value represented by the sum of the atomic weights of all the atoms in the molecule. For example, the MW of ethyl-dichloroarsine, $C_2H_5AsCl_2$, is as follows:

C	Atomic weight =	12.0	× 2 =	24.0
H	Atomic weight =	1.0	× 5 =	5.0
As	Atomic weight =	74.9	× 1 =	74.9
Cl	Atomic weight =	35.5	× 2 =	71.0
	Molecular weight =			174.9

High MWs tend to indicate solids. Conversely, low MWs tend to indicate gases. Agents easily broken down by heat often have very high MWs. These very high MWs may indicate that decontamination by fire is practical.

The MW gives an indication of the persistency of an agent. Generally, the higher the number, the lower the rate of evaporation and the greater the persistency. Another use for the MW is in calculating the vapor density.

MW indicates the ability of a chemical to penetrate filters. MWs of 29 or less are very difficult to stop with activated charcoal filters. Examples are ammonia (NH_3, MW=17) and carbon monoxide (CO, MW=28).

Rate of Detoxification

The human body can detoxify some toxic materials. This rate of detoxification is the rate at which the body can counteract the effects of a poisonous substance. It is an important factor in determining the hazards of repeated exposure to toxic chemical agents.

Most chemical agents are essentially cumulative in their effects. The reason is that the human body detoxifies them very slowly or not at all. For example, a 1-h exposure to HD or CG followed within a few hours by another 1-h exposure has about the same effect as a single 2-h exposure. Continued exposure to low concentrations of HD may cause sensitivity to very low concentrations of HD. Other chemical agents also have cumulative effects. For example, an initial exposure to a small (less than lethal) amount of Sarin (GB) would decrease cholinesterase levels; a second quantity less than the LD_{50}

could be enough to kill. (Although the body can detoxify it to some extent, GB is essentially cumulative.)

Some compounds have a detoxification rate that is significant. Because the body detoxifies such chemical agents as AC and cyanogen chloride (CK) at a fairly rapid rate, it takes high concentrations of these agents to produce maximum casualty effects.

Rate of Action

The rate of action of a chemical agent is the rate at which the body reacts to or is affected by that agent. The rate varies widely, even to those of similar tactical or physiological classification. For example, blister agent HD causes no immediate sensation on the skin. Skin effects usually occur several hours later (some cases result in delays of 10–12 days before symptoms appear). In contrast, lewisite produces an immediate burning sensation on the skin upon contact and blistering in about 13 h. Decontamination immediately (within 4–5 min) will prevent serious blister agent effects.

With the single exception of arsine (SA), the nerve agents and the blood agents are very fast acting. Vomiting compounds also exert their effects within a short time after inhalation. In general, agents that are inhaled or ingested will affect the body more rapidly than those that contact the skin. To avert death, first aid measures, such as administering antidotes, generally must follow within a few minutes after the absorption of a lethal dosage of any agents.

Vapor Density

Vapor density is the ratio of the density of any gas or vapor to the density of air, under the same conditions of temperature and pressure. It is a measure of how heavy the vapor is in relation to the same volume of air. Vapor density helps in estimating how long an agent will persist in valleys and depressions. The higher the vapor density, the longer the vapor will linger in low-lying areas.

To calculate the vapor density, divide the MW by 29 (the average MW of air). If the result is greater than 1.0, the agent is heavier than air. The agent will tend to collect in low-lying areas, such as foxholes and ditches, and in vehicles. If the result is less than 1.0, the agent almost invariably is nonpersistent. It will quickly dissipate into the atmosphere.

For example, phosgene, $COCl_2$, with an MW of 98.92, has a vapor density of 3.4× that of air. The calculation follows:

$$\frac{phosgene}{air} \quad \frac{98.92}{29} = 3.4 \quad \frac{3.4}{1} = 3.4$$

Phosgene will persist 5 min or longer in the open in the summer. Hydrogen cyanide, MW=27.02, may persist only a minute under the same circumstances.

Agents with low vapor densities rise. Agent AC is the only militarily significant agent that is lighter than air. Chemical agent clouds with high vapor densities seek lower ground in much the same way as water poured on the ground. Agents with high vapor densities tend to evolve into long coherent streamers. For example, agent GB is almost 5× as heavy as air. It can produce a streamer 40 m wide by 22 km long from a single round.

Diffusion is usually a minor factor in the dissemination of chemical agents. This is especially so after air dilutes the chemical agent. Air currents and other influences tend to offset any effects of diffusion or vapor density. Because very high vapor densities cause agents to seek lower ground, the agent can overcome the effects of wind to a limited degree. These densities can actually cause the agent cloud to go upwind if the upwind area is lower than the downwind area and the wind is not very strong. This explains the upwind radius on all chemical hazard predictions.

Vapor Pressure

Vapor pressure is the pressure exerted by a vapor when a state of equilibrium exists between the vapor and its liquid (or solid) state. It is the pressure in a closed space above a substance when no other gas is present. Vapor pressure varies with temperature, so the temperature must be stated to determine vapor pressure. At any temperature any liquid (or solid) will have some vapor pressure, however small.

Substances with high vapor pressure evaporate rapidly. Those with low vapor pressure evaporate slowly. The impact of vapor pressure on the rate of evaporation makes vapor pressure a very important property in considering the tactical use and duration of effectiveness of chemical agents. A potential chemical agent is valuable for employment when it has a reasonable vapor pressure. One with exceptionally high vapor pressure is of limited use. It vaporizes and dissipates too quickly. Examples are arsine and carbon monoxide. On the other hand, mechanical or thermal means may effectively aerosolize and disseminate solid and liquid agents of very low vapor pressure. Vapor pressure and volatility are related. Translated into volatility, vapor pressure is most understandable and useful.

Volatility

Volatility is the weight of vapor present in a unit volume of air, under equilibrium conditions, at a specified temperature. It is a measure of how much material (agent) evaporates under given conditions. The volatility depends on vapor pressure. It varies directly with temperature. We express volatility as milligrams of vapor per cubic meter (mg/m^3). Calculate it numerically by an equation derived from the perfect gas law. This equation follows:

$$V = \frac{16020 \times MW \times VP}{K}$$

where

V = volatility,

MW = molecular weight,

VP = vapor pressure in mm Hg at a specified temperature,

K = Kelvin temperature = Celsius temperature + 273.

You need to know more than the vapor pressure or volatility to judge the effectiveness of a chemical agent. You must also consider the degree of toxicity of physiological action of the chemical agent. A highly toxic chemical agent of relatively low volatility, such as GB, may be far more lethal than a less toxic chemical agent of much higher volatility, such as CG.

Appendix C

Chemical Warfare Agents, Signs, and Syndromes

Agents	Signs	Symptoms	Onset	Clinical diagnostic tests	Exposure route and treatment
Nerve Agents: Sarin (GB) Tabun (GA) Soman (GD) Cyclosarin (GF) VX Novichok agents, and other organophosphorus compounds including carbamates and pesticides.	Pinpoint pupils (miosis), bronchoconstriction, respiratory arrest, hypersalivation, increased secretions, diarrhea, decreased memory, concentration, loss of consciousness, and seizures	**Moderate exposure:** diffuse muscle cramping, runny nose, difficulty in breathing, eye pain, dimming of vision, sweating, and muscle tremors. **High exposure:** the above plus sudden loss of consciousness, seizures, and flaccid paralysis (late sign).	**Aerosols:** seconds to minutes. **Liquids:** minutes to hours.	Red blood cell or serum cholinesterase (whole blood). Treat based on signs and symptoms; lab tests only for later confirmation.	**Inhalation and dermal absorption:** Atropine (2 mg) IV; repeat q 5 min, titrate until effective, average dose 6 to >15 mg – use IM in the field before IV access (establish airway for oxygenation). Pralidoxime chloride (2-PAMCl) 600–1800 mg IM or 1.0 g IV over 20–30 min (maximum 2 g IM or IV per hour). Additional doses of atropine and 2-PAMCl depending on severity. Diazepam or lorazepam to prevent seizures if >4 mg atropine given ventilatory support.

Agent	Signs and Symptoms		Time of Onset	Diagnosis	Treatment
Cyanides: hydrogen cyanide (HCN) and cyanogen chloride	**Moderate exposure:** metabolic acidosis, venous blood-O_2 level above normal, hypotension, "pink" skin color. **High exposure:** above signs plus coma, convulsions, cessation of respiration, and heartbeat.	**Moderate exposure:** giddiness, palpitations, dizziness, nausea, vomiting, headache, eye irritation, increase in rate and depth of breathing (hyperventilation), and drowsiness. **High exposure:** immediate loss of consciousness, convulsions and death within 1–15 min.	Seconds to minutes.	Bitter almond odor associated with patient suggests cyanide poisoning, metabolic acidosis, cyanide (blood) or thiocyanate (blood or urine) levels. Treat based on signs and symptoms; lab tests only for later confirmation.	**Inhalation, ingestion and dermal absorption:** 100% oxygen by facemask; intubation with 100% FiO_2 if indicated. Amyl nitrite through inhalation, 1 ampule (0.2 mL) q 5 min. Sodium nitrite (300 mg IV over 5–10 min) and sodium thiosulfate (12.5 g IV). Additional sodium nitrite should be based on hemoglobin level and weight of patient.
Vesicants/blister agents: sulfur mustard, lewisite, nitrogen mustard, mustard lewisite, and phosgene oxime.	Skin erythema and blistering; watery, swollen eyes; upper airways sloughing with pulmonary edema; metabolic failure; neutropenia and sepsis (especially sulfur mustard, late in course).	Burning, itching, or red skin, mucosal irritation (prominent tearing, and burning and redness of eyes), shortness of breath, nausea, and vomiting.	Lewisite, minutes; Sulfur mustard, hours to days	Often smell of garlic, horseradish, and/or mustard on body. Oily droplets on skin from ambient sources. Urine thiodiglycol. Tissue biopsy (USAMRICD).	**Inhalation and dermal absorption:** mustards no antidote. For lewisite and ewisite/mustard lmixtures: British Anti-Lewisite (BAL or Dimercaprol) IM (rarely available). Thermal burn therapy; supportive care (respiratory support and eye care).

(Continued)

Agents	Signs	Symptoms	Onset	Clinical diagnostic tests	Exposure route and treatment
Pulmonary/choking agents: phosgene, chlorine, diphosgene, chloropicrin, oxides of nitrogen, and sulfur dioxide.	Pulmonary edema with some mucosal irritation (greater water solubility of agent is equal to greater mucosal irritation) leading to ARDS or non-cardiogenic pulmonary edema. Pulmonary infiltrate.	Shortness of breath, chest tightness, wheezing, laryngeal spasm, mucosal and dermal irritation, and redness.	1–24 h (rarely up to 72 h); may be asymptom atic period of hours.	No tests available, but history may help to identify source and exposure characteristics (majority of incidents generating exposures to humans involve trucking with labels on vehicle).	**Inhalation, ingestion and dermal adsorption:** no antidote. Management antidote. Management of secretions; O_2 therapy; consider high dose steroids to prevent pulmonary edema (demonstrated benefit only for oxides of nitrogen). Treat pulmonary edema with PEEP to maintain PO_2 above 60 mm Hg.

Appendix D

FTIR/MS Spectra

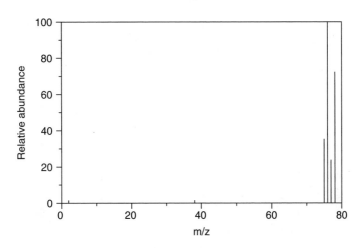

Figure D 1. Mass spectrum of arsine – SA

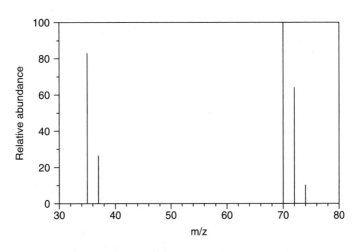

Figure D 2. Mass spectrum of chlorine – Cl

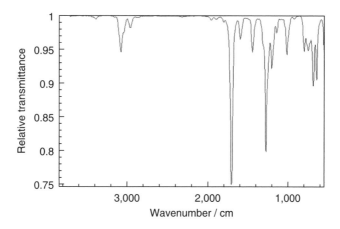

Figure D 3. FTIR spectrum of chloroacetophenone – CN

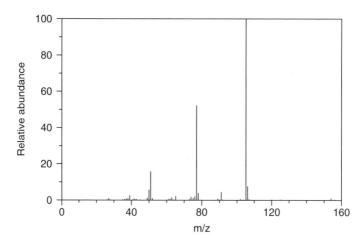

Figure D 4. Mass spectrum of chloroacetophenone – CN

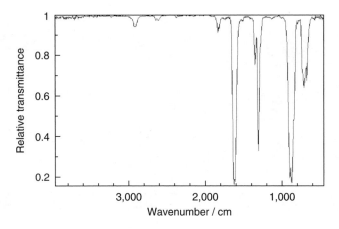

Figure D 5. FTIR spectrum of chloropicrin – PS

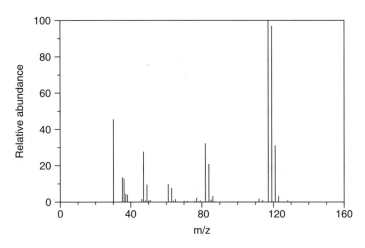

Figure D 6. Mass spectrum of chloropicrin – PS

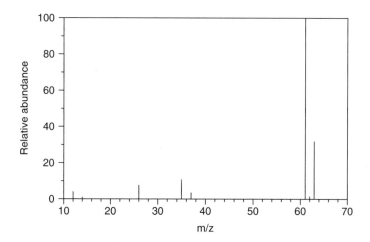

Figure D 7. Mass spectrum of cyanogen chloride – CK

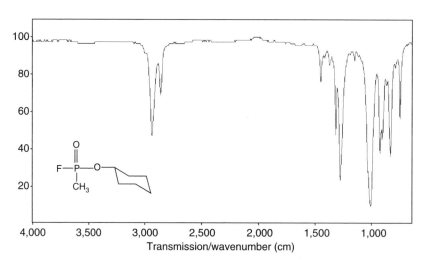

Figure D 8. FTIR spectrum of cyclosarin – GF

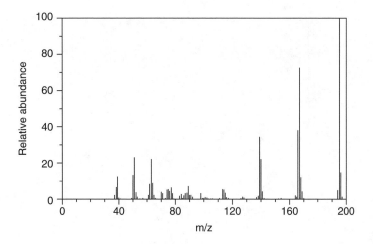

Figure D 9. Mass spectrum of dibenzoxazepine – CR

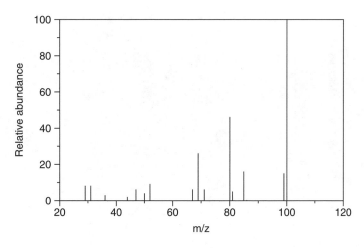

Figure D 10. Mass spectrum of difluoro – DF

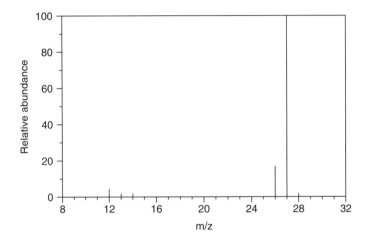

Figure D 11. Mass spectrum of hydrogen cyanide – AC

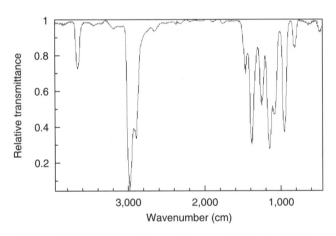

Figure D 12. FTIR spectrum of isopropyl alcohol – OPA-1

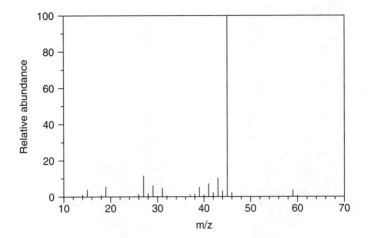

Figure D 13. Mass spectrum of isopropyl alcohol – OPA-1

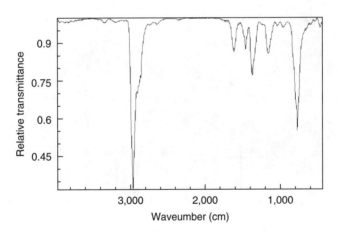

Figure D 14. FTIR spectrum of isopropylamine – OPA-2

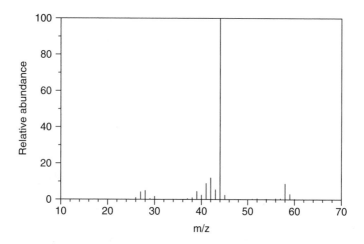

Figure D 15. Mass spectrum of isopropylamine – OPA-2

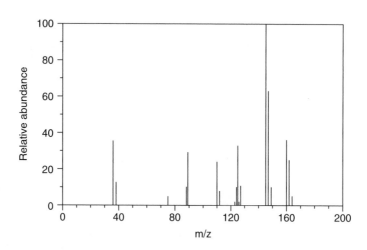

Figure D 16. Mass spectrum of methyldichloroarsine – MD

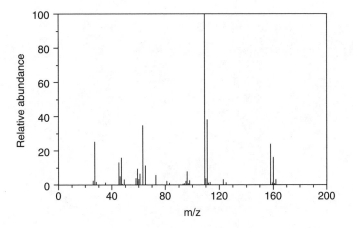

Figure D 17. Mass spectrum of mustard – HD

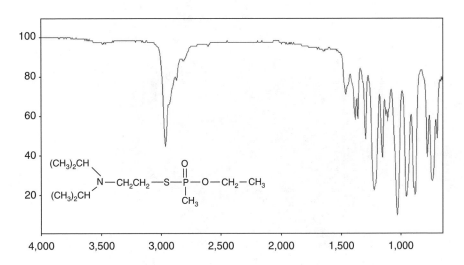

Figure D 18. FTIR spectrum of nerve agent VX

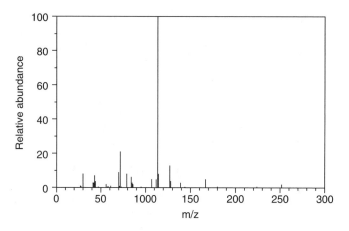

Figure D 19. Mass spectrum of nerve agent VX

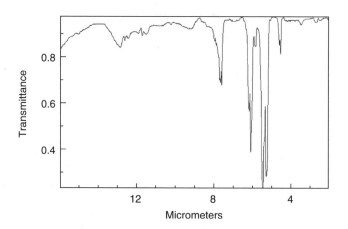

Figure D 20. FTIR spectrum of nitric oxide – NO

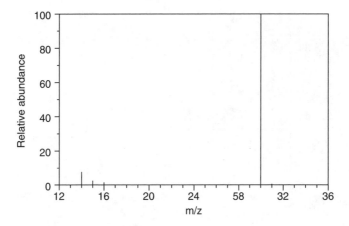

Figure D 21. Mass spectrum of nitric oxide – NO

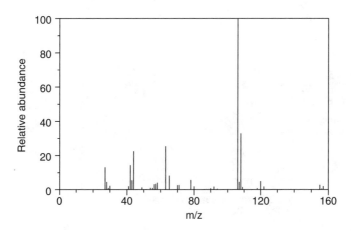

Figure D 22. Mass spectrum of nitrogen mustard 2 – HN-2

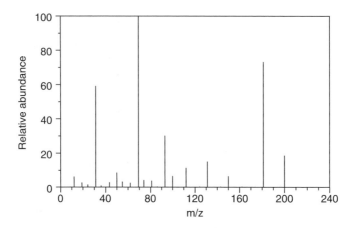

Figure D 23. Mass spectrum of perfluoroisobutylene – PFIB

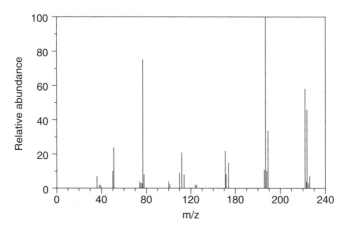

Figure D 24. Mass spectrum of phenyldichloroarsine – PD

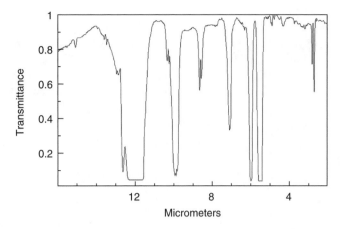

Figure D 25. FTIR spectrum of phosgene – CG

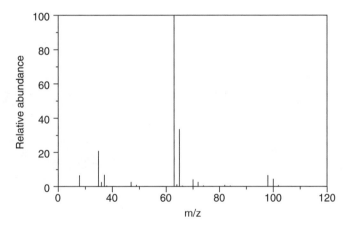

Figure D 26. Mass spectrum of phosgene – CG

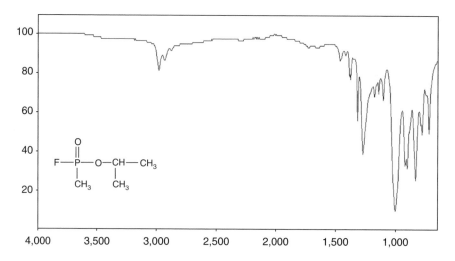

Figure D 27. FTIR spectrum of sarin – GB

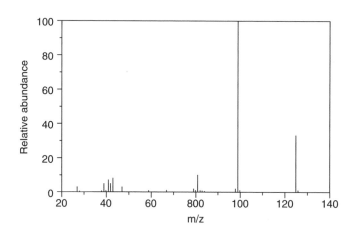

Figure D 28. Mass spectrum of sarin – GB

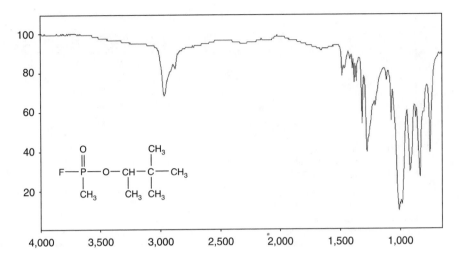

Figure D 29. FTIR spectrum of soman – GD

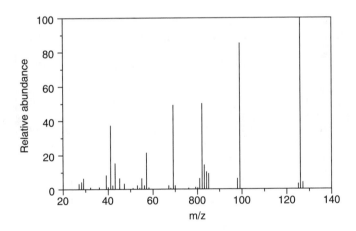

Figure D 30. Mass spectrum of soman – GD

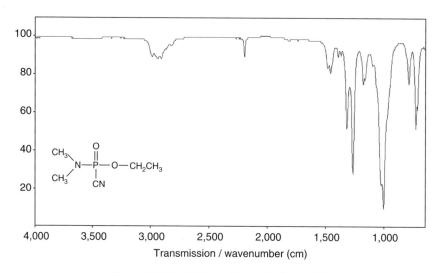

Figure D 31. FTIR spectrum of tabun – GA

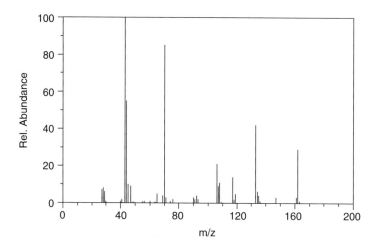

Figure D 32. Mass spectrum of tabun – GA

Appendix E

Chemical Warfare Agent Cross-Reference Table

Table E 1. Chemical warfare agent cross-reference table

Agent	Symbol	Type	CAS number
Adamsite	DM	Vomit	[578-94-9]
Agent BZ	BZ	Incapacitating	[6581-06-2]
Alcohol – amine mixture	OPA	Binary	[Not available]
Isopropyl alcohol			[67-63-0]
Isopropyl amine			[75-31-0]
Amiton	VG	Nerve	[78-53-5]
Arsine	SA	Blood	[7784-42-1]
Binary component	QL	Binary	[57856-11-80]
Bromobenzylcyanide	CA	Tear	[5798-79-8]
Chlorine	CL	Choking	[7782-50-5]
Chloroacetophenone	CN	Tear	[532-27-4]
Chloropicrin	PS	Tear	[76-06-2]
Cyanogen chloride	CK	Blood	[506-77-4]
Cyclosarin	GF	Nerve	[329-99-7]
Difluoro	DF	Binary	[676-99-3]
Diphenylchloroarsine	DA	Vomit	[712-48-1]
Diphenylcyanoarsine	DC	Vomit	[23525-22-6]
Diphosgene	DP	Choking	[503-38-8]
Distilled sulfur mustard	HD	Blister	[505-60-2]
Edemo	VM	Nerve	[21770-86-5]
Ethyl sarin	GE	Nerve	[1189-87-3]
Ethyldichloroarsine	ED	Blister	[598-14-1]
Hydrogen cyanide	AC	Blood	[74-90-8]
Lewisite 1	L1	Blister	[541-25-3]
Lewisite 2	L2	Blister	[40334-69-8]
Lewisite 3	L3	Blister	[40334-70-1]
Methyldichloroarsine	MD	Blister	[593-89-5]
Mustard – lewisite mixture	HL	Blister	[Not available]
Sulfur mustard	HD		[505-60-2]
Lewisite 1	L1		[541-25-3]
Mustard T	T	Blister	[63918-89-8]
Nerve agent – GV	GV	Nerve	[141102-74-1]

(Continued)

209

Table E 1. Chemical warfare agent cross-reference table—Cont'd

Agent	Symbol	Type	CAS number
Nerve agent – VE	VE	Nerve	[21738-25-0]
Nerve agent – VS	VS	Nerve	[73835-17-3]
Nerve agent – Vx	V_X	Nerve	[20820-80-8]
Nerve agent – VX	VX	Nerve	[50782-69-9]
Nitric oxide	NO	Choking	[10102-43-9]
Nitrogen mustard 1	HN-1	Blister	[538-07-8]
Nitrogen mustard 2	HN-2	Blister	[51-75-2]
Nitrogen mustard 3	HN-3	Blister	[555-77-1]
o-Chlorobenzylidene malononitrile	CS	Tear	[2698-41-1]
Perfluoroisobutylene	PFIB	Choking	[382-21-8]
Phenyldichloroarsine	PD	Blister	[696-28-6]
Phosgene	CG	Choking	[75-44-5]
Phosgene oxime	CX	Blister	[1794-86-1]
Russian VX	VR	Nerve	[159939-87-4]
Sarin	GB	Nerve	[107-44-8]
Sesquimustard	Q	Blister	[3563-36-8]
Soman	GD	Nerve	[96-64-0]
Sulfur mustard mixture	HT	Blister	[172672-28-5]
Sulfur mustard	HD		[505-60-2]
Mustard T	T		[63918-89-8]
Tabun	GA	Nerve	[77-81-6]
Tear agent – CNB	CNB	Tear	[Not available]
Chloroacetophenone			[532-27-4]
Carbon tetrachloride			[56-23-5]
Benzene			[71-43-2]
Tear agent – CNC	CNC	Tear	[Not available]
Chloroacetophenone			[532-27-4]
Chloroform			[67-66-4]
Tear agent – CNS	CNS	Tear	[Not available]
Chloroacetophenone			[532-27-4]
Chloropicrin			[76-06-2]
Chloroform			[67-66-3]
Tear agent – CR	CR	Tear	[257-07-8]

Appendix F

Chemical Warfare Agent Precursor Chemicals: Uses and Equivalents

Table F 1. Chemical warfare agent precursor chemicals: uses and equivalents

Precursor chemical CAS registry number	Civilian uses	CW agent production	Units of agent per unit of precursor[a]
1. Thiodiglygol 111-48-8	Organic synthesis Carrier for dyes in textile industry Lubricant additives Manufacturing plastics	Sulfur mustard (HD) Sesquimustard (Q)	1.3 1.79
2. Phosphorus oxychloride 10025-87-3	Organic synthesis Plasticizers Gasoline additives Hydraulic fluids Insecticides Dopant for semi-conductor grade silicon Flame retardants	Tabun (GA)	1.05
3. Dimethyl methyl phosphonate (DMMP) 756-79-6	Flame retardant	Sarin (GB) Soman (GD) Cyclosarin (GF)	1.12 1.45
4. Methylphos-phonyl difluoride 676-99-3	Organic synthesis Specific uses not identified	Sarin (GB) Soman (GD) Cyclosarin (GF)	1.40 1.82 1.80
5. Methylphos-phonyl dichloride 676-97-1	Organic synthesis Specific uses not identified	Sarin (GB) Soman (GD) Cyclosarin (GF)	1.05 1.36 1.35
6. Dimethyl phosphite 868-85-9	Organic synthesis Specific uses not identified	Sarin (GB) Soman (GD) Cyclosarin (GF)	1.27 1.65 1.65

(Continued)

211

Table F 1. Chemical warfare agent precursor chemicals: uses and equivalents—Cont'd

Precursor chemical CAS registry number	Civilian uses	CW agent production	Units of agent per unit of precursor[a]
7. Phosphorus trichloride 7719-12-2	Organic synthesis Insecticides Gasoline additives Plasticizers Surfactants	Amiton (VG) Tabun (GA) Sarin (GB) Salt process Rearrangement process	1.95 1.18 1.02 (0.34)[b] 1.02/(0.68)[b]
	Dyestuffs	Soman (GD) Salt process Rearrangement process Cyclosarin (GF) Salt process Rearrangement process	1.32 (0.44)[b] 1.32/(0.88)[b] 1.31 (0.44)[b] 1.31/(0.87)[b]
8. Trimethyl phosphite 121-45-9	Organic synthesis	Used to make dimethyl methyl phosphonate (DMMP) – molecular rearrangement	See dimethyl methyl phosphonate
9. Thionyl chloride[b] 7719-09-7 Could serve as chlorinating agent in all of these processes – other chlorinating agents could be substituted	Organic synthesis Chlorinating agent Catalyst Pesticides Engineering plastics	Sarin (GB) Soman (GD) Cyclosarin (GF) Sulfur mustard (HD) Sesquimustard (Q) Nitrogen mustard (HN-1) Nitrogen mustard (HN-2) Nitrogen mustard (HN-3)	1.18 1.53 1.51 1.34 1.84 0.714 0.655 1.145
10. 3-Hydroxy-1-methylpiperidine 3554-74-3	Specific uses not identified; probably used in pharmaceutical industry	Nonidentified; could be used in the synthesis of psychoactive compounds such as BZ	
11. N,N-diisopropyl-(beta)-amino-ethyl chloride 96-79-7	Organic synthesis	VX VS	1.64 1.72
12. N,N-diisopropyl-aminoethane-thiol 5842-07-9	Organic synthesis	VX VS	1.66 1.75

13. 3-Quinuclidinol 1619-34-7	Hypotensive agent Probably used in synthesis of pharmaceuticals	BZ	2.65
14. Potassium fluoride 7789-23-3	Fluorination of organic compounds Cleaning and disinfecting brewery, dairy, and other food processing equipment Glass and porcelain manufacturing	Sarin (GB) Soman (GD) Cyclosarin (GF)	2.41 3.14 3.10
15. 2-Chloroethanol 107-07-3	Organic synthesis Manufacturing of ethylene oxide and ethylene glycol Insecticides Solvent	Sulfur mustard (HD) Sesquimustard (Q) Nitrogen mustard (HN-1)	0.99 0.99 1.06
16. Dimethylamine 124-40-3	Organic synthesis Pharmaceuticals Detergents Pesticides Gasoline additive Missile fuels Vulcanization of rubber	Tabun (GA)	3.61
17. Diethyl ethylphos- phonate 78-38-6	Heavy metal extraction Gasoline additive Antifoam agent Plasticizer	Ethyl sarin (GE)	0.93
18. Diethyl *N,N*- dimethyl phosphoramidate 2404-03-7	Organic synthesis Specific uses not identified	Tabun (GA)	0.90
19. Diethyl phosphite 762-04-9	Organic synthesis Paint solvent Lubricant additive	Amiton (VG) Sarin (GB) Soman (GD) Cyclosarin (GF)	Catalyst 1.02 1.32 1.30
20. Dimethylamine HCl 506-59-2	Organic synthesis Pharmaceuticals Surfactants Pesticides Gasoline additives	Tabun (GA)	1.99

(Continued)

Table F 1. Chemical warfare agent precursor chemicals: uses and equivalents—Cont'd

Precursor chemical CAS registry number	Civilian uses	CW agent production	Units of agent per unit of precursor[a]
21. Ethylphos- phonous dichloride 1498-40-4	Organic synthesis Specific uses not identified but could be used in manufacturing of flame retardants, gas additives, pesticides, and surfactants	VE VS Ethyl sarin (GE)	1.93 2.14 1.18
22. Ethylphosphonyl dichloride 1066-50-8	Organic synthesis Specific uses not identified; see ethylphosphonous dichloride	Ethyl sarin (GE)	2.10
23. Ethylphosphonyl difluoride 753-98-0	Organic synthesis Specific uses not identified; see ethylphosphonous dichloride	Ethyl sarin (GE)	2.70
24. Hydrogen fluoride 7664-39-3	Fluorinating agent in chemical reactions Catalyst in alkylation and polymerization reactions Additives to liquid rocket fuels Uranium refining	Sarin (GB) Soman (GD) Ethyl sarin (GE) Cyclosarin (GF)	7.0 9.11 7.7 9.01
25. Methyl benzilate 76-89-1	Organic synthesis Tranquilizers	BZ	1.39
26. Methylphos- phonous dichloride 676-83-5	Organic synthesis	VX	2.28
27. N,N-diisopropyl- (beta)- aminoethanol 96-80-0	Organic synthesis Specific uses not identified	VX	1.84
28. Pinacolyl alcohol 464-07-3	Specific uses not identified	Soman (GD)	1.79

29. *O*-ethyl, 2-diisopropyl aminoethyl methylphos-phonate (QL) 57856-11-8	Specific uses not identified	Specific uses not identified Organic synthesis	1.14 Amiton (VG)
30. Triethyl phosphite 122-52-1	Organic synthesis Plasticizers Lubricant additives	Amiton (VG)	1.62
31. Arsenic trichloride 7784-34-1	Organic synthesis Pharmaceuticals Insecticides Ceramics	Arsine (SA) Lewisite (L) Adamsite (DM) Diphenylchlo-roarsine (DA)	0.43 1.14 1.53 1.45
32. Benzilic acid 76-93-7	Organic synthesis	BZ	1.48
33. Diethyl methyl-phosphonite 15715-41-0	Organic synthesis	VX	1.97
34. Dimethyl ethyl-phosphonate 6163-75-3	Organic synthesis	Ethyl sarin (GE)	1.12
35. Ethylphos-phonous difluoride 430-78-4	Organic synthesis	VE Ethyl sarin (GE)	2.58 1.57
36. Methylphos-phonous difluoride 753-59-3	Organic synthesis	VX VM Sarin (GB) Soman (GD) Cyclosarin (GF)	3.18 2.84 1.67 2.17 2.15
37. 3-Quinuclidone 1619-34-7	Same as 3-Quinuclidinol	BZ	2.65
38. Phosphorous pentachloride 10026-13-8	Organic synthesis Pesticides Plastics	Tabun (GA)	0.78
39. Pinacolone 75-97-8	Specific uses not identified	Soman (GD)	1.82
40. Potassium cyanide 151-50-8	Extraction of gold and silver from ores Pesticide Fumigant Electroplating	Tabun (GA) Hydrogen cyanide (AC)	1.25 0.41

(Continued)

Table F 1. Chemical warfare agent precursor chemicals: uses and equivalents—Cont'd

Precursor chemical CAS registry number	Civilian uses	CW agent production	Units of agent per unit of precursor[a]
41. Potassium bifluoride 7789-29-9	Fluorine production Catalyst in alkylation Treatment of coal to reduce slag formation	Sarin (GB) Soman (GD) Cyclosarin (GF)	1.79 2.33 2.31
42. Ammonium bifluoride 1341-49-7	Ceramics Disinfectant for food equipment Electroplating Etching glass	Sarin (GB) Soman (GD) Cyclosarin (GF)	2.46 3.20 3.16
43. Sodium fluoride 7681-49-4	Pesticide Disinfectant Dental prophylaxis Glass and steel manufacturing	Sarin (GB) Soman (GD) Cyclosarin (GF)	3.33 4.34 4.29
44. Sodium bifluoride 1333-83-1	Antiseptic Neutralizer in laundry operations Tinplate production	Sarin (GB) Soman (GD) Cyclosarin (GF)	2.26 2.94 2.91
45. Sodium cyanide 143-33-9	Extraction from gold and silver ores Fumigant Manufacturing dyes and pigments Core hardening of metals Nylon production	Tabun (GA) Hydrogen cyanide (AC) Cyanogen chloride (CK)	1.65 0.55 1.25
46. Triethanolamine 102-71-6	Organic synthesis Detergents Cosmetics Corrosion inhibitor Plasticizer Rubber accelerator	Organic synthesis Detergents Cosmetics Corrosion inhibitor Plasticizer Rubber accelerator	1.37
47. Phosphorus pentasulfide 1314-80-3	Organic synthesis Insecticide Mitocides Lubricant oil additives Pyrotechnics	Amiton (VG) VX	1.21 1.20
48. Diisopropy-lamine 108-18-9	Organic synthesis Specific uses not identified	VX	3.65

49. Diethylamino-ethanol 100-37-8	Organic synthesis Anticorrosion composition Pharmaceuticals Textile softeners	Amiton (VG) VM	2.30 2.05
50. Sodium sulfide 1313-82-2	Paper manufacturing Rubber manufacturing Metal refining Dye manufacturing	Sulfur mustard (HD)	2.04
51. Sulfur monochloride and sulfur chloride 10025-67-9	Organic synthesis Pharmaceuticals Sulfur dyes Insecticides Rubber vulcanization Polymerization catalyst Hardening of soft woods Extraction of gold from ores	Sulfur mustard (HD)	1.18
52. Sulfur dichloride 10545-99-0	Organic synthesis Rubber vulcanizing Insecticides Vulcanizing oils Chlorinating agents	Sulfur mustard (HD)	1.54
53. Triethanolamine hydrochloride 637-39-8	Organic synthesis Insecticides Surface active agents Waxes and polishes Textile specialties Lubricants Toiletries Cement additive Petroleum demulsifier Synthetic resin	Nitrogen mustard (HN)	1.10
54. N,N-diisopropyl-2-aminoethyl chloride hydrochloride 4261-68-1	Organic synthesis	VX	1.34

a Assumes quantitative reaction yields.
b Figures in parentheses are based on the use of PCl_3 as a chlorine donor in the reaction.

Appendix G

Periodic Table of the Elements

PERIODIC TABLE OF THE ELEMENTS

GROUP NUMBERS
IUPAC RECOMMENDATION (1985)

GROUP NUMBERS
CHEMICAL ABSTRACT SERVICE (1986)

ATOMIC NUMBER — 5   10.811 — RELATIVE ATOMIC MASS (1)
SYMBOL — B
BORON — ELEMENT NAME

PERIOD	1 IA	2 IIA	3 IIIB	4 IVB	5 VB	6 VIB	7 VIIB	8 VIIIB	9 VIIIB	10 VIIIB	11 IB	12 IIB	13 IIIA	14 IVA	15 VA	16 VIA	17 VIIA	18 VIIIA
1	1 1.0079 H HYDROGEN																	2 4.0026 He HELIUM
2	3 6.941 Li LITHIUM	4 9.0122 Be BERYLLIUM											5 10.811 B BORON	6 12.011 C CARBON	7 14.007 N NITROGEN	8 15.999 O OXYGEN	9 18.998 F FLUORINE	10 20.180 Ne NEON
3	11 22.990 Na SODIUM	12 24.305 Mg MAGNESIUM											13 26.982 Al ALUMINIUM	14 28.086 Si SILICON	15 30.974 P PHOSPHORUS	16 32.065 S SULPHUR	17 35.453 Cl CHLORINE	18 39.948 Ar ARGON
4	19 39.098 K POTASSIUM	20 40.078 Ca CALCIUM	21 44.956 Sc SCANDIUM	22 47.867 Ti TITANIUM	23 50.942 V VANADIUM	24 51.996 Cr CHROMIUM	25 54.938 Mn MANGANESE	26 55.845 Fe IRON	27 58.933 Co COBALT	28 58.693 Ni NICKEL	29 63.546 Cu COPPER	30 65.39 Zn ZINC	31 69.723 Ga GALLIUM	32 72.64 Ge GERMANIUM	33 74.922 As ARSENIC	34 78.96 Se SELENIUM	35 79.904 Br BROMINE	36 83.80 Kr KRYPTON
5	37 85.468 Rb RUBIDIUM	38 87.62 Sr STRONTIUM	39 88.906 Y YTTRIUM	40 91.224 Zr ZIRCONIUM	41 92.906 Nb NIOBIUM	42 95.94 Mo MOLYBDENUM	43 (98) Tc TECHNETIUM	44 101.07 Ru RUTHENIUM	45 102.91 Rh RHODIUM	46 106.42 Pd PALLADIUM	47 107.87 Ag SILVER	48 112.41 Cd CADMIUM	49 114.82 In INDIUM	50 118.71 Sn TIN	51 121.76 Sb ANTIMONY	52 127.60 Te TELLURIUM	53 126.90 I IODINE	54 131.29 Xe XENON
6	55 132.91 Cs CAESIUM	56 137.33 Ba BARIUM	57-71 La-Lu Lanthanide	72 178.49 Hf HAFNIUM	73 180.95 Ta TANTALUM	74 183.84 W TUNGSTEN	75 186.21 Re RHENIUM	76 190.23 Os OSMIUM	77 192.22 Ir IRIDIUM	78 195.08 Pt PLATINUM	79 196.97 Au GOLD	80 200.59 Hg MERCURY	81 204.38 Tl THALLIUM	82 207.2 Pb LEAD	83 208.98 Bi BISMUTH	84 (209) Po POLONIUM	85 (210) At ASTATINE	86 (222) Rn RADON
7	87 (223) Fr FRANCIUM	88 (226) Ra RADIUM	89-103 Ac-Lr Actinide	104 (261) Rf RUTHERFORDIUM	105 (262) Db DUBNIUM	106 (266) Sg SEABORGIUM	107 (264) Bh BOHRIUM	108 (277) Hs HASSIUM	109 (268) Mt MEITNERIUM	110 (281) Uun UNUNNILIUM	111 (272) Uuu UNUNUNIUM	112 (285) Uub UNUNBIUM	114 (289) Uuq UNUNQUADIUM					

LANTHANIDE

57 138.91 La LANTHANUM	58 140.12 Ce CERIUM	59 140.91 Pr PRASEODYMIUM	60 144.24 Nd NEODYMIUM	61 (145) Pm PROMETHIUM	62 150.36 Sm SAMARIUM	63 151.96 Eu EUROPIUM	64 157.25 Gd GADOLINIUM	65 158.93 Tb TERBIUM	66 162.50 Dy DYSPROSIUM	67 164.93 Ho HOLMIUM	68 167.26 Er ERBIUM	69 168.93 Tm THULIUM	70 173.04 Yb YTTERBIUM	71 174.97 Lu LUTETIUM

ACTINIDE

89 (227) Ac ACTINIUM	90 232.04 Th THORIUM	91 231.04 Pa PROTACTINIUM	92 238.03 U URANIUM	93 (237) Np NEPTUNIUM	94 (244) Pu PLUTONIUM	95 (243) Am AMERICIUM	96 (247) Cm CURIUM	97 (247) Bk BERKELIUM	98 (251) Cf CALIFORNIUM	99 (252) Es EINSTEINIUM	100 (257) Fm FERMIUM	101 (258) Md MENDELEVIUM	102 (259) No NOBELIUM	103 (262) Lr LAWRENCIUM

(1) Pure Appl. Chem. 73, No. 4, 667-683 (2001)

Relative atomic mass is shown with five significant figures. For elements have no stable nuclide, the value enclosed in brackets indicates the mass number of the longest-lived isotope of the element.

However three such elements (Th, Pa, and U) do have a characteristic terrestrial isotopic composition, and for these an atomic weight is tabulated.

Index

Printed in the United States of America.